The Niger River Basin: A Vision for Sustainable Management

The Niger River Basin: A Vision for Sustainable Management

J. ROBERT VAN PELT LIBRARY
MICHIGAN TECHNOLOGICAL UNIVERSITY
HOUGHTON, MICHIGAN

Inger Andersen
Ousmane Dione
Martha Jarosewich-Holder
Jean-Claude Olivry

Edited by Katherin George Golitzen

THE WORLD BANK
Washington, DC

© 2005 The International Bank for Reconstruction and Development / The World Bank
1818 H Street, NW
Washington, DC 20433
Telephone: 202-473-1000
Internet: www.worldbank.org
E-mail: feedback@worldbank.org

All rights reserved.

1 2 3 4 08 07 06 05

The findings, interpretations, and conclusions expressed in this paper do not necessarily reflect the views of the Executive Directors of The World Bank or the governments they represent.

The World Bank does not guarantee the accuracy of the data included in this work. The boundaries, colors, denominations, and other information shown on any map in this work do not imply any judgment on the part of The World Bank concerning the legal status of any territory or the endorsement or acceptance of such boundaries.

Rights and Permissions

The material in this publication is copyrighted. Copying and/or transmitting portions or all of this work without permission may be a violation of applicable law. The International Bank for Reconstruction and Development / The World Bank encourages dissemination of its work and will normally grant permission to reproduce portions of the work promptly.

For permission to photocopy or reprint any part of this work, please send a request with complete information to the Copyright Clearance Center Inc., 222 Rosewood Drive, Danvers, MA 01923, USA; telephone: 978-750-8400; fax: 978-750-4470; Internet: www.copyright.com.

All other queries on rights and licenses, including subsidiary rights, should be addressed to the Office of the Publisher, The World Bank, 1818 H Street, NW, Washington, DC 20433, USA; fax: 202-522-2422; e-mail: pubrights@worldbank.org.

ISBN-10: 0-8213-6203-8
ISBN-13: 978-0-8213-6203-7
e-ISBN: 0-8213-6204-6
DOI: 10.1596/978-0-8213-6203-7

Library of Congress Cataloging-in-Publication Data has been applied for.

Contents

Foreword vii

Acknowledgments ix

Executive Summary x

Acronyms and Abbreviations xv

**1 Overview of Niger River Basin Countries
and Basin History** 1
The Countries of the Niger River Basin 4
History of the Basin 7

2 The Physical Geography of the Niger River Basin 11
Physical Environment and Hydrography 11
Navigable River Segments 19
Geology and Hydrogeology 20
Soils 22
Natural Environment 24
Climatic Conditions 25
Climate and Water Resources Variability 27

3 The Niger River Basin's Water Resources 30
Hydrology 30
Transport of Suspended and Dissolved Solids 48
Water Quality 56
Closing Comments on the Technical Chapters 57

**4 Cooperative Development of the Niger River Basin:
Criteria for Success** 58
Promoting Development and Poverty Reduction 58
From Unilateral to Cooperative Development 59
Laying the Institutional Foundation for Cooperation 60

A Political Mandate: The Shared Vision and Sustainable
    Development Action Program 63
Making Cooperation Happen in the Niger River Basin 63
People and the Environment: A Focus of Cooperation 65
Criteria for Success and Ways Forward 68

Appendixes **70**
Appendix 1: Main Maps of the Niger River Basin 70
Appendix 2: Technical Supporting Information 86
Appendix 3: Overview of Data Management 127
Appendix 4: Glossary 129

Endnotes **131**

Bibliography **132**

Index **141**

Foreword

The Niger River Basin, home to approximately 100 million people, is a vital, complex asset for West and Central Africa. It is the continent's third longest river (4,200 kilometers), traversing nine countries—Benin, Burkina Faso, Cameroon, Chad, Côte d'Ivoire, Guinea, Mali, Niger, and Nigeria. The Niger River embodies the livelihoods and geopolitics of the nations it crosses. This river is not simply water, but is also an origin of identity, a route for migration and commerce, a source of potential conflict, and a catalyst for cooperation.

As the regional population and economies grow, this life-giving resource will require integrated water resources management to address sustainably the increasing aspirations and needs of the people of the Basin. Regional cooperation among the nations sharing the Basin is crucial to optimizing sustainable economic development while ensuring environmentally sound management. Over previous decades, the Niger River Basin countries have focused more on unilateral water resources development than on the potential benefits of cooperation. Recently, however, the Basin countries have renewed their commitment to address sustainable management and development of the Basin's resources through an improved framework and a Shared Vision process. In doing so, they are setting an important path for cooperative water resources planning, management, and development that could serve as a model for water resources management for Africa's many shared rivers.

From the Guinea highlands, through the Office du Niger in the Inland Delta, to the Niger Delta, the greatest opportunities will come from managing efficiently the water resources on which millions of people depend for their livelihoods, taking into consideration the range of demands, associated infrastructures, and ecological complexity. As in all river basins, different groups of users have conflicting priorities and preferences. Thus, it is important to understand how the river and its resources are used in order to avoid conflict and promote cooperation.

This overview of the potential of the Niger River's water resources discusses the hydrographic system—the river and its tributaries, hydrology, climate, and water quality and use. It attempts to capture the full spectrum of the Niger ecosystem's values and benefits and to support the integration of science and decision making, and, as such, can serve as a tool for transboundary cooperative management of shared water resources. The final chapter discusses the criteria for success that will be required to broker agreement for the cooperative development and management of the river. The book is presented as a contribution to support and foster the Shared Vision process for integrated management and development of this majestic river, for the benefit of the people of the Niger River Basin.

Michel Wormser
Sector Director
Private Sector and Infrastructure
Africa Region

Acknowledgments

This book intends to contribute to the Niger River Basin Shared Vision process. Having devoted much of his academic career to the study of the Niger River, Jean-Claude Olivry has produced an extensive study, based on his own research and a compilation of existing information. The study by Jean-Claude Olivry forms the background for this book. Drawing from his work, a team comprising Ousmane Dione, Martha Jarosewich-Holder, and Katherin Golitzen prepared chapters 1–3. The concluding chapter has been written by Inger Andersen (Director, Water, Environment, Social and Rural Development Unit, Middle East and North Africa Region), with input from the World Bank team, and provides a discussion of the criteria for success in water resources development in the Basin.

The team would like to thank the Niger Basin Authority for its cooperation and support, namely Mohammed Bello Tuga (Executive Secretary), Ousmane Diallo (Environmental Specialist), Oumar Ould Ali (Chief Hydrologist), Robert Dessoussi (Hydrologist), and the Niger Basin Authority Focal Points within the nine member countries. The team also extends its thanks for reviews, input, suggestions, and comments received from Johan Grijsen, Amal Talbi, Esther Monier-Illouz, and cartographer Jeff Lecksell. The team is grateful for the technical input and peer review of Alessandro Palmieri, Lead Dam Specialist, Environmentally and Socially Sustainable Development Network, World Bank. In addition, the team has benefited greatly from advice and guidance provided by David Grey, Senior Water Adviser, Africa Region, World Bank. Finally, the team expresses its appreciation to the government of the Netherlands for funding the translation and publication through a grant from the Bank-Netherlands Water Partnership Program.

Executive Summary

This book comprises two distinct elements. The first, and major, part of the book (chapters 1–3) is a unique and essential compilation of technical information and data on the entire Niger River system. It presents a comprehensive overview of the physical environment and hydrological functions of the watershed, thus providing the necessary background for examination of the challenges of resource management and development potential. The second part of the book (chapter 4) presents the fundamental challenges that the nine countries[1] of the Basin face and are now addressing.

Introduction

The river is described in the first three chapters as a single hydrologic system, yet it has largely been developed unilaterally, in spite of the existence of a convention and a river basin organization. Chapter 4 presents a description of the path that the political leaders of the Basin have chosen to take and a discussion of what is needed to achieve their goals. It is the combination of these elements—the hydrology and geography of the river system with the economic opportunities, institutions, politics, and diplomacy—that is essential to achieve joint and optimal development of this shared resource.

The Niger River Basin, host to approximately 100 million people, is a large catchment area with significant water resources development potential. Effective development presupposes international and bilateral agreements, together with national commitment to sustainable use of shared resources. Such commitment is particularly important given that the many variations in geography, culture, demographics, and economics among the Basin countries lead to differing needs and expectations with regard to the Basin's resources. Coordination and cooperation among decision makers and users is crucial to address the threats to water resources from natural causes and human activities. The Niger Basin Authority (NBA) is mandated to foster this cooperation and sharing of resources at the national and regional levels. The NBA brings together the Basin countries to understand the

EXECUTIVE SUMMARY   xi

complex interrelated dynamics of the Niger River Basin and to promote integrated water resources management.

To this end, the Basin countries have agreed that the Niger Basin Shared Vision will be implemented through a Sustainable Development Action Program (SDAP). Through a multidisciplinary dialogue, the SDAP will contribute to building the institutional capacity, policy harmonization, and public support that will create the framework for enhanced cooperation among the Basin countries.

Overview of Niger River Basin Countries and Basin History

The Niger River Basin is an extraordinary asset for the nine countries that are within its watershed and for the broader West and Central Africa region. Each country within the Basin has unique geographic settings and a wide range of available resources. Chapter 1 summarizes country geographic settings and socioeconomic characteristics. The countries can be clustered as "water resources producers"—Guinea, Cameroon, and to a lesser extent Benin; or "water resources consumers"—Mali and Niger. Nigeria is both a producer and a consumer. Côte d'Ivoire, Burkina Faso, and Chad are part of the Basin but are minimally affected by the use and management of the river's water resources.

Historical water resources use, from sustenance to commerce, has influenced the evolution of development and cooperation in the region. The European colonial legacy of exploration and expansion was followed by the independence movement in the 1960s, in which an initial effort was made to define the conditions for shared development of the Basin. These efforts evolved over 20 years, and in 1980, the present NBA replaced the Niger River Commission. The evolution of the NBA has continued to the present. Its leadership has a clear commitment to the integrated management of the Basin's environmental and water resources.

Physical Geography of the Niger River Basin

The Niger River is Africa's third longest river (4,200 kilometers) and encompasses six hydrographic regions, each of which is distinguished by unique topographic and drainage characteristics. The Upper Niger River Basin headwaters are in the Fouta Djallon Massif, Guinea. From there the river flows northeast, traversing the Inland Delta, a vast spreading floodplain (averaging 50,000 square kilometers) that dissipates an appreciable portion of its potential hydraulics through absorption and evaporation. When it reaches the fringes of the Sahara Desert, the Niger River turns back by forming a great bend and flowing south and east as the Middle Niger River section, then as the Lower Niger, to the Niger Delta at the Gulf of Guinea, which it reaches after being joined by its largest tributary, the Benue River.

The diverse geographic and climatic characteristics of the Niger River Basin play an important role in water resources availability, which in turn affects a range of water resources–related activities. The Niger River water system is one of the most impressive examples of the influence of topography and climate on the flow conditions of a water system. Such a large basin area cannot be expected to have uniform climatic and rain patterns, and the Niger River traverses a wide range of ecosystem zones in West Africa. A combination of human population growth, unsustained resource use and development, and desertification threatens the Niger River's ability to supply crucially needed natural resources to the people of the Basin. The geology and soils of the Basin also influence groundwater availability. Significant rainwater deficits and the variable duration of the rainy season result in hydrological deficits that are not necessarily reflected in a direct response of the base flow. Chapter 2 reviews the Basin's physical geography, six hydrographic regions, geology, groundwater and soils, natural environment, and climatic characteristics.

The Niger River Basin's Water Resources

The Basin is a unique and complex river system with an extensive network of tributaries. Using available information, chapter 3 evaluates the Basin's water resources in the six corresponding hydrologic reaches. Because of climatic variations, the annual river flood does not occur at the same time in different parts of the Basin. There are usually high flows from the headwaters in Guinea, a decrease in flow caused by evaporation and expansion in the floodplain of the Inland Delta, followed by an increase in flow from tributary input through the Middle and Lower reaches as the river enters the Niger Delta. In the Upper Niger, the high-water discharges generally occur in September, and the low-water season is generally April–May. The Inland Delta has an estimated storage capacity of 70 cubic kilometers but has a high rate of loss caused by evaporation over the thousands of square kilometers of its floodplain. This loss is estimated at about 44 percent of the inflow. The peak flow period that arrives in September is delayed as it spreads out, exiting the Inland Delta three months later. A phase of receding water extends into February. In the Middle Niger, at Niamey, the maximum flows are usually twofold: a first wet seasonal peak flow and the upstream peak flow that arrives during the dry season. The first high-water discharge, known as the white flood (because of the light sediment content of the water), occurs soon after the rainy season in September. A second rise, known as the black flood, begins in December with the arrival of inflow from upstream. May and June are the low-water months in the Middle Niger.

On the Benue, there is only one high-water season, because of the Benue's more southerly climatic location; this normally occurs from May to October, which is earlier than on the Middle Niger. The Lower Niger below its confluence with the Benue consequently has a high-water period that begins in May or June and a low-water period that is at least a month shorter than on the Middle Niger, because the rains in the south start earlier. In terms of water quality, an increase in siltation is linked to erosion, deforestation, and soil depletion.

Cooperative Development of the Niger River Basin— Criteria for Success

A better understanding of the Niger River Basin will assist decision makers in basin management. It is a premise of river basin management that managing the river as a system yields optimal benefits. In the case of the Niger River, this could mean increased water, food, power, transport, and so on. Optimized management of any river is difficult, primarily caused by the need to recognize so many different interests. Management of an international river is particularly difficult, but much can still be done to move toward optimized management.

Once cooperative investments have been made in the development of the water resources, trust and cooperation will grow between the countries and many other benefits will accrue, including those "beyond the river," such as communication investments, increases in trade, improved flows of labor, and so on, thereby leading to better regional integration of the countries of the basin. Specific investment opportunities identified by the countries include enhanced food and energy production; transportation; environmental management, such as investments in land productivity and measures against desertification; flood and drought management; and investments in livestock, fisheries, and tourism.

The NBA Summit of Heads of State has set the organization on a renewed path, through the Shared Vision process and SDAP. If the NBA is to succeed in revitalizing itself so that it can drive regional development of the river, several criteria for success will be required of the institution, its stakeholders, and the donor community. These success criteria are necessary to ensure that the reengagement and renewal, which are currently taking place within the NBA, will take hold.

Institutionally, the NBA will need to earn and recapture legitimacy, relevance, and support from its constituency. National engagement from governments and other key stakeholders—in the form of a strong champion and an adequate coordination mechanism for river basin management—is critical to moving development forward. For national water resources management and development aspirations to be fulfilled by the shared water resources, a broad national constituency must have ownership of the agenda. The degree to which the NBA can recapture both legitimacy and relevance will largely determine whether the institution will meet the expectations of its constituency. This is all the more important because NBA financial sustainability, which is key to its renewal and survival, will be secure only after the constituency sees the relevance and benefits from the institution.

The Niger Basin Summit of Heads of State has embarked on a Shared Vision process. This is a bold commitment, moving from a past of unilateral actions on the river toward enhanced coordination, collaboration, and joint action. The process is an expression of the political commitment of the heads of state to a cooperative agenda. The Shared Vision will guide the formulation of the SDAP, which will identify and define the development opportunities in which the Basin countries can jointly participate. The Shared Vision and the SDAP will form a platform for mobilizing resources from the NBA countries and from the donor communities for investments to implement the SDAP.

To succeed in moving this process forward, the NBA will need to continue toward greater transparency, inclusivity, and engagement of the communities and stakeholders who live with and on the river. Issues such as escalating populations, conflict and war, and environmental stresses will continue to put increased pressure on the river and its resources. Although the NBA cannot address all these issues, the organization can be an important platform for awareness of transboundary impacts of socioeconomic pressures on natural resources. The subsidiarity principle will help the NBA, as part of the SDAP, to identify areas where the institution will have a comparative advantage over well-established national and local agencies, which are also charged with working on these matters.

The path ahead is clearly difficult. As the countries move forward, the key ingredients for success include continued strong political leadership and champions, staying the course of the reform process, maintaining a dynamic and enabled staff, and sustaining a financially viable institution that continues to stay on message, to move beyond unilateral planning, to facilitate hydrodiplomacy, and to engage donors to commit to their side of the compact.

Acronyms and Abbreviations

AGRHYMET	Centre Régional de Formation et d'Application de la Météorologie et de l'Hydrologie Opérationnelle (Regional Center for Training and Application of Meteorology and Operational Hydrology)
AOC-HYCOS	Afrique de l'Ouest et Centrale (West and Central Africa) Hydrological Cycle Observing System
CIDA	Canadian International Development Agency
CILSS	Comité Permanent Inter-Etats de Lutte contre la Sécheresse au Sahel (Permanent Interstate Committee for Drought Control in the Sahel)
CIP	Centre Inter-Etats de Prévision (Interstate Forecasting Center)
FONDAS	Niger River Basin Development Fund
GDP	gross domestic product
GNI	gross national income
GHENIS	Gestion hydro-écologique du Niger supérieur Mali-Guinée
HA	hydrological areas
HDI	Human Development Index
HYDRONIGER	Hydrological Forecasting System in the Niger River Basin
IRD	Institut de Recherche pour le Développement (Research Institute for Development; formerly ORSTOM)
JICA	Japan International Cooperation Agency
NBA	Niger Basin Authority
NOAA-AVHRR	United States National Oceanic and Atmospheric Administration (Advanced Very High Resolution Radiometer)
OMVS	Organisation pour la Mise en Valeur du Fleure Senegal (Senegal River Basin Organization)

ORSTOM	Office pour la Recherche Scientifique et Technique d'Outre Mer (Office for Overseas Scientific and Technical Research; now IRD)
PIRT	Projet d'Inventaire des Ressources Terrestres
SDAP	Sustainable Development Action Program
TDS	total dissolved solids
TOC	total organic carbon
Ts	transport spécifique (specific transport)
TSS	total suspended solids
UN	United Nations
UNDP	United Nations Development Programme
UNESCO	United Nations Educational, Scientific, and Cultural Organization
WDI	*World Development Indicators*
WHO-Oncho	World Health Organization (Onchocerciasis Control Program in West Africa)
WHYCOS	World Hydrological Cycle Observing System
WMO	World Meteorological Organization

1
Overview of Niger River Basin Countries and Basin History

The Niger River is shared by nine countries in West and Central Africa—Benin, Burkina Faso, Cameroon, Chad, Côte d'Ivoire, Guinea, Mali, Niger, and Nigeria (see appendix 1, map 1). The nine Basin countries are among the poorest in the world. Four are among the bottom 20 countries on the *World Development Indicators* (WDI) scale,[2] and seven are among the bottom 20 on the United Nations Development Programme (UNDP) Human Development Index (HDI).[3] The need for development and investment in the region is evident, and the Niger River holds tremendous development potential. Development opportunities range from those directly related to the river, such as power, irrigation, and navigation, to those "beyond the river," such as increases in trade, communication investments, and enhanced labor flows.

The Basin's contrasting geographic settings and variable availability of raw materials and natural resources drive local and regional development. The Niger River's hydrologically active basin covers a surface area of nearly 1.5 million square kilometers shared among the nine countries according to the following approximate percentages: Benin (2.5 percent), Burkina Faso (3.9 percent), Cameroon (4.4 percent), Chad (1.0 percent), Côte d'Ivoire (1.2 percent), Guinea (4.6 percent), Mali (30.3 percent), Niger (23.8 percent), and Nigeria (28.3 percent). The population living in the Basin is estimated at 100 million, with an average growth rate of 3 percent per year. In terms of managing and using the Niger's water resources, the nine Basin countries can be clustered as "water resources producers"—Guinea, Cameroon, and to a lesser extent Benin; or "water resources consumers"—Mali and Niger. Nigeria is both a producer and a consumer. Côte d'Ivoire, Burkina Faso, and Chad are part of the Basin but are minimally affected by the use and management of the river's water resources. Table 1.1 summarizes the general statistical socioeconomic characteristics of the Niger Basin countries and their relative characteristics as part of the Basin, and describes the individual socioeconomic setting of each country. Maps 2–8 in appendix 1 illustrate the countries' geographic features and Basin characteristics.

Table 1.1 Socioeconomic Statistical Characteristics of the Niger River Basin Countries

Parameter	Benin	Burkina Faso	Cameroon	Chad	Côte d'Ivoire	Guinea	Mali	Niger	Nigeria
Total area (millions km^2)	0.114	0.274	0.475	1.284	0.322	0.246	1.24	1.27	0.924
Population (millions)	6.75	10.7	14.9	8.3	15.4	7.1	10.6	10.7	114
Population increase (%/year)	3.1	2.3	2.3	3.2	2.1	3.1	2.2	3.5	2.7
Urban population (%)	39.9	18	48.1	23.5	45.8	32.1	29.4	20.1	43.1
GDP/person (US$)	933	965	1,573	850	1,653	1,934	753	753	853
Estimated population 2025 (millions)	11.5	21.7	27.8	13.4	29.9	14	22.7	19.2	235
Agricultural production (1,000 tons)									
Rice	36	89	65	100	1,162	750	590	54	3,400
Peanuts	—	205	160	372	144	182	140	—	2,783
Corn	662	378	600–850	173	573	89	341	5	5,127
Millet	29	979	71	366	65	10	641	2,391	5,956
Sugar cane	—	—	—	280	—	220	303	174	675
Cotton	150	136	75–79	103	130	16	218	—	55

Livestock (head, millions)									
Beef	1.35	4.55	5.90	5.58	1.35	2.37	6.06	2.17	*19.8*
Sheep	0.63	6.35	3.80	2.43	1.39	0.69	*6.0*	*4.31*	*20.5*
Freshwater fishing (1,000 tons)	44	—	89	6	68	103	108	6	383
Within Niger Basin									
Hydrologically active Area (10³ km²)[a]	37.50	58.5	66.0	15.0	18.0	69.0	454.50	357.0	424.50
(%)	2.50	3.90	4.40	1.0	1.20	4.60	30.30	23.80	28.30
Population (millions)	1.95	2.12	4.46	0.08	0.80	1.60	7.80	8.30	67.60
(%)	2.10	2.20	4.70	0.10	0.80	1.70	8.20	8.80	71.40

Source: Data are primarily based on recent (2004) national multisector studies on the assessment of opportunities and constraints to development of each country's portion of the Niger River Basin, prepared with support from the World Bank (Cameroon, Chad, Guinea, and Nigeria) and the Canadian International Development Agency (CIDA) (Benin, Burkina Faso, Côte d'Ivoire, Mali, and Niger).

Note: — = not available. Values in italics relate to data involving the entire country; data in plain text relate to the part of the country located in the Niger River Basin.

a. http://www.riob-info.org/gwp/PP_Niger.pdf. The hydrologically active area of the Niger River Basin is 1.5 million square kilometers.

The Countries of the Niger River Basin

Benin

The catchments of several right-bank tributaries of the Middle Niger River are situated in northern Benin, occupying 2.5 percent of the total area of the Basin (37,500 square kilometers). Benin is densely populated, with 65 inhabitants per square kilometer, on average. More than 1.95 million people live in the Niger Basin in Benin. The land within the Basin is used primarily for grazing and livestock, although there are areas, once used for groundnut farming, that are now used for cotton farming. Cotton farming in this area now contributes one-third of the national production. The main city is Kandi, and Malanville is the river port. The railroad from Cotonou reaches only to Parakou, in the center of the country, thus limiting access for commerce to Kandi. The Mekrou River, a tributary of the Niger, crosses the "W" International Park, an extensive protected sanctuary for flora and fauna, shared by Benin, Burkina Faso, and Niger.

Burkina Faso

Similar to Benin, several tributaries of the Niger River originate in Burkina Faso. About 25 percent (58,500 square kilometers) of the country's total territory is within the Basin, comprising about 3.9 percent of the Basin's area. One-quarter of Burkina Faso's population, 2.12 million, lives in the Basin, which is the driest region of the country. It is also the poorest economic area, with pastoral livelihoods that are seminomadic in the north and sedentary in the south.

Cameroon

The headwaters of the Benue River, a major tributary of the Niger River, lie within Cameroon, comprising 4.4 percent (66,000 square kilometers) of the Basin. A population of 4.46 million lives in this part of the Basin. The Benue watershed is predominantly agricultural, producing cotton and peanuts for export. Livestock is also important, especially on the Adamawa Plateau. High expectations for floodplain irrigation development have not been met, despite the construction of the Lagdo Dam. This dam was also expected to add a few additional months of navigation on the Benue, establishing Garoua as a river port through which cotton could be shipped downstream, to be traded for manufactured products and raw materials. However, continued low water flows have limited the duration of the navigation period. The potential hydropower production from the Lagdo Dam, however, is significantly greater than the needs of the immediate region, providing an opportunity to export electricity. Farther downstream, areas irrigated by the Bamenda and Nkambe tributaries of the Metchum River support high-altitude grazing, in addition to tea, coffee, and corn production that add to the revenues of a densely populated area. Paddy rice is grown in the lowlands. Faro, Benue, and Boubandjida national parks are protected areas that ensure the conservation of flora and fauna and protection of the headwater areas of the upper Benue River.

Chad

The headwaters of the Mayo tributaries lie within Chad, comprising 1.0 percent (15,000 square kilometers) of the Basin. A population of 80,000 lives in this part of the Basin, extending through the upper Mayo Kebi and Kabia tributaries and the lakes of the Toubouris depression. The Basin supports a sparsely settled rural population that depends on subsistence farming and limited cotton production for its livelihood.

Côte d'Ivoire

The headwaters of the Bagoé, the Kankelaba, and the Baoulé tributaries lie within Côte d'Ivoire, comprising 1.2 percent (18,000 square kilometers) of the Basin. A population of 800,000 lives in this part of the Basin. The area is a southern extension of the cotton-growing areas of Mali. It also produces kola nuts and is, above all, an area for raising livestock and a route for transporting cattle herds to the coast and to Abidjan. Locally, small development projects (ponds and small dams) provide opportunities for market gardening and aquaculture. The main cities are Odienne and Boundiali.

Guinea

The Guinea highlands (Haute Guinée) and part of the high plateaus of Guinée Forestière are two geographic provinces in Guinea situated at the headwaters of the Niger River and its tributaries, comprising 4.6 percent (69,000 square kilometers) of the Basin. About 1.6 million people live in this part of the Basin, more than 80 percent of whom live in rural areas, with a density of fewer than 30 inhabitants per square kilometer. The alluvial valleys of these two provinces are fertile, yielding most of the country's agricultural production. The highlands and plateaus have fewer than five inhabitants per square kilometer, and are dedicated to raising livestock and limited rainfed farming, generally of sorghum and millet. Rainfall is sufficient for groundnut production in the Tinkisso watershed (about 60,000 tons per year, which is about one-third of the production in Guinea), and paddy rice production in the valleys of the Niger, the Niandan, and the Milo Rivers (estimated at 300,000 tons per year). Coffee is also grown (20,000 tons per year) on the mountain ridges and in the upper Milo River region. These two geographic provinces are rich in diamonds. Bauxite deposits are mined on the Tinkisso and gold is mined close to Siguiri.

Mali

The longest reach of the Niger River (1,700 kilometers) extends through southern Mali, comprising 30.3 percent of the Basin (454,500 square kilometers). A population of 7.8 million lives in the Basin, with a large percentage living in the capital, Bamako, that is situated on the river. This city, with its industrial and suburban areas and its market gardens, plays a significant role in the economic development in this part of the Basin. The Office du Niger, one of the largest and oldest irrigation schemes in West Africa, continues to provide opportunities for agricultural development. Irrigated farming along the river produces 590,000 tons of rice and 303,000 tons of sugar cane per year. The Inland Delta is an undeveloped, flooded ecosystem with abundant freshwater fishing areas, high productivity pastureland, and fertile agricultural lands. Catches from freshwater fishing average about 108,000 tons per year, with an additional 10,000 tons harvested on the Selingue reservoir, depending on the water level and flooding in the Inland Delta. Its excellent pastureland makes the Inland Delta a grazing area for more than 2 million head of cattle. Cotton production covers an area of 122,000 square kilometers. On a yearly basis, Mali produces 400,000 tons of cottonseed and 218,000 tons of cotton fiber, surpassing Egypt as the largest producer in Africa. Whereas it is an irrigated crop in Egypt, it is a rainfed crop in Mali. Southern Mali also produces more than 230,000 tons of millet and sorghum, 215,000 tons of corn, and 18,000 tons of groundnuts per year, all as rainfed crops.

Niger

The river's reach through Niger is 540 kilometers, with a hydrologically active area of approximately 357,000 square kilometers occupying proportionally 23.8 percent of the Basin, with a population of 8.3 million. This area extends through the Maradi region, which is part of the Sokoto watershed. The left-bank tributary network, originating in the Aïr and the Azaouâk Mountains, is characterized by intermittent flows that are isolated from the

Niger River without any hydrologic connection to the river network. The Niger River partially irrigates large alluvial plains and lowlands of the Dallol Bosso, the Dallol Maouri, and the Maradi area. Rice production is low, but production of traditional grains of the Sahel region, although subject to climate variations, is significant (more than 2.4 million tons). In many places, black-eyed peas have replaced groundnuts as an export crop and cotton is no longer grown. Agricultural practices have undergone significant changes since the severe droughts of the past several decades. Niger is dependent on the navigable waterways of the Niger River through Nigeria.

Nigeria

Nigeria is the final downstream country through which the Niger River flows, and contains 28.3 percent (424,500 square kilometers) of the Basin area. The Niger Basin extends across 20 of the 36 states of Nigeria and comprises two main rivers, the Niger and the Benue, and 20 tributaries. Of Nigeria's major rivers, more than half are in the Niger River Basin. Their combined length accounts for almost 60 percent of the total length of all important rivers in Nigeria. Almost 60 percent of Nigeria's population, or about 67.6 million inhabitants, live in the Basin. These Nigerians comprise 80 percent of the population of the entire Basin. Given Nigeria's size and location, its agricultural production, both rainfed and irrigated, is substantial (see table 1.1). Nigeria is also the largest oil-producing country on the African continent and the sixth largest in the world.

History of the Basin

A brief look at the history of the Niger River and the Basin provides an understanding of its present-day role in Central and West Africa. Throughout history, the river has played an essential role in the lives of the people who depend on it for their livelihoods.

Historical Use

The Niger River's original name, *egerou n-igereou* (river of rivers), was given by the Tuareg people to express the exceptional character they attributed to it. Over the centuries, the Niger River and its tributaries have been vital to those who lived along the river or used it for travel and trade. For travelers coming from the Sahara, the appearance of the Inland Delta as the sea of fresh water was a welcome relief. Riverbank sediment was used to make mud bricks for homes. Men and boys fished on the river daily, individually or as part of groups in the Inland Delta organized by and under the supervision of the Maître des Eaux (Master of the Waters) who determined what was to be done on the river, and where and how it was to be done, according to the flood level, especially during flood recession. The sequence of flooding in the Basin allowed for flow-recession agriculture after the flood season and dry cropping along the banks and in the marshes. With regard to commerce, large Nigerien fishing vessels brought agricultural products from the south to Timbuktu and returned with salt, which came from the Sahara by camel caravan from Taoudenni. Travel on the river linked communities of the Basin. Precolonial armies and, later, the colonial military followed the travel routes taken by early travelers and merchants.

From the 11th century, four great nations dominated in the vast grasslands and highlands of the Basin in West Africa, trading in gold and salt: the empires of Ghana (11th–13th centuries), Mali (13th–15th centuries), Songhai (14th–16th centuries), and Kamen–Borno (16th–17th centuries). In the coastal areas, there were two prominent kingdoms: the Kingdom of Yoruba (12th–14th centuries) and the Kingdom of Benin (15th–17th centuries). During the 16th and 17th centuries, trade with Europeans increased along the West African coast, leading to European domination and colonization in the region. In appendix 1, map 9 is a historical map depicting the knowledge of the Niger Basin in 1830.

Colonial Era

The first agreements signed by the colonial powers regarding the Niger River region concerned the division of territory among France, Germany, and Great Britain.[4] The initial treaty among the three powers, signed at the Conference of Berlin on February 26, 1885, proclaimed complete freedom of navigation and commerce based on equality among the colonial powers. Germany used the Benue to reach northern Cameroon; navigation on the Middle Niger had not yet been developed. Other treaties on commerce were also enacted during this period, such as the General Act and Declaration of Brussels in 1890. With the Franco-German treaty of 1911, France ceded to Germany the territories to the east and south of the colony of Cameroon and, notably, the eastern part of the Benue watershed (formerly Mayo Kebi), now in Chad. After Germany's defeat in World War I, the Agreement of Saint-Germain-en-Laye, signed on September 10, 1919, was an order by the League of Nations handing over the former German colony of Cameroon to France and England, and renewing the provisions concerning navigation on the Niger and the Benue. No further changes in colonial borders took place after that date and the Entente Cordiale (Cordial Agreement) among the colonizing countries allowed effective use of the navigable waterways.

Postcolonial Era

During the independence movement in the 1960s, the above-mentioned treaties were abrogated and, in the spirit of free commerce, the nine Basin countries established joint customs houses and instituted taxes on imported goods. Beginning in February 1960, a working session between Mali and Nigeria defined the conditions for shared development of the Niger River Basin, with support from the president of Niger. The meeting of the 16th session of the Commission for Technical Cooperation in Sub-Saharan Africa, at Mamou, Guinea, in 1960, recommended cooperation and exchange of technical information and data among Basin hydrologists, although this did not occur immediately. The First Conference on the Niger River Basin in which all nine Basin countries participated was held in May 1961, at Segou, Mali. This first step in fostering cooperation among the Basin water users underscored the importance of coordination and the dangers of undertaking national projects on a unilateral basis. As the newly independent countries became members of the United Nations (UN), they also agreed to comply with UN international conventions, such as the Barcelona Convention of 1921 concerning the use and regime of international navigable waterways.

The Second Conference on the Niger River Basin, held at Niamey in October 1963, resulted in a new act, signed by the nine Basin countries, which renewed navigational freedom on the Niger and its tributaries. It also introduced the principles for cooperation necessary to evaluate and implement development projects that could affect the river system.[5] This act set the stage for the establishment of an intergovernmental organization for cooperation. The Niger River Commission was established at a meeting held at Niamey on November 25, 1964. The agreement establishing the commission defined new rules concerning agricultural and industrial use of water, water resources development, navigation, and transportation. The agreement was amended in 1968 and again in 1973 (Godana 1985).

In 1980, the heads of state of the Basin countries signed a new convention to create the Niger Basin Authority (NBA), replacing the Niger River Commission. The text of this convention amended the agreement of 1964 and amendments of 1968 and 1973. The convention stated that the NBA's long-term objective was to promote cooperation between and among the member states and to ensure integrated development in all areas as part of development of its resources, particularly in the areas of energy, water, agriculture, forestry, transportation and communication, and industry. At the same time, it recognized each country's individual right to exploit resources within its portion of the Basin. A protocol signed in Faranah, Guinea, in November 1980, created the Niger River Basin Development Fund, referred to as FONDAS. The purpose of FONDAS was to have a financial mechanism to support the goals of the NBA. Despite this transition from commission to authority, which was intended to bolster the organization's institutional effectiveness and operational self-sufficiency, the lack of financial commitment on the part of member countries led to a gradual loss of credibility. Consequently, several development partners withdrew support.[6]

This situation prevailed until the 17th ordinary session of the NBA in Abuja, Nigeria, in 1998, when the ministers of the member states met to discuss this institutional problem and identify a mechanism to address systematically the progressive degradation of the environment and water resources of the Basin. In February 2002, the nine heads of state met in Abuja to renew their political commitment to manage the Basin's water resources in a sustainable manner and to maximize development opportunities through multilateral cooperation. The NBA's heads of state agreed to launch the Shared Vision for the Basin's sustainable development, supported by the SDAP. Following this political commitment, the member states commissioned an institutional audit of the NBA to strengthen its institutional capacity at the NBA's Extraordinary Council of Ministers meeting in Yaoundé, Cameroon, in January 2004. Following the Yaoundé meeting, the nine heads of state signed the Paris Declaration in April 2004, confirming the member states' commitment to the Niger Basin Shared Vision process.[7] This process is further discussed in chapter 4.

2
The Physical Geography of the Niger River Basin

Physical Environment and Hydrography

The diverse geographic and climatic characteristics of the Niger River Basin play an important role in water resources availability, which in turn affects a range of water resources–related activities. The Niger River water system is one of the most impressive examples of the influence of topography and climate on the flow conditions of a water system. Such a large basin area cannot be expected to have uniform climatic and rain patterns, and the Niger River traverses almost all the possible ecosystem zones in West and Central Africa. A combination of human population growth, unsustained resource use and development, and desertification threatens the Niger River's ability to supply crucially needed natural resources to the people of West Africa. The geology and soils of the Basin also influence groundwater availability. Significant rainwater deficits and the variable duration of the rainy season result in hydrological deficits that are not necessarily reflected in a direct response of the base flow. This chapter reviews the Basin's physical geography, six hydrographic regions, geology, groundwater and soils, natural environment, and climatic characteristics.

Physical Environment

The Niger River Basin is located between the meridians of 11°30' west and 15° east, from Guinea to Chad; and between the parallels of 22° north and 5° north, from the Hoggar Mountains to the Gulf of Guinea. The Basin extends 3,000 kilometers from east to west and 2,000 kilometers from north to south. Originating in the Guinean highlands within the regions of Haute Guinée and Guinée Forestière in the Fouta Djallon Massif, the Niger River is the third-longest river in Africa (4,200 kilometers), after the Nile and the Congo. The Upper Niger River takes its source from the Fouta Djallon Massif in Guinea at an altitude of about 800 meters and flows northeast, traversing the Inland Delta in Mali. The Inland Delta is a vast, spreading floodplain (averaging 40,000 square kilometers) that dissipates a significant proportion of the flow of the river through absorption and evaporation.

When it reaches the fringes of the Sahara Desert, the Niger River turns back by forming a great bend and flowing south and east as the Middle Niger River section, then as the Lower Niger to the Niger Delta at the Gulf of Guinea. Before reaching the Niger Delta, the Niger River is joined by its major tributary, the Benue River, which originates in the highlands of Cameroon's Adamawa Plateau. The full extent of the Basin has been the subject of debate, because there is no input to the Niger River system from its left bank as it runs through Niger. From the headwaters to the Niger Delta, therefore, taking into account the hydrologically active area, the Basin has an average area of about 1.5 million square kilometers.

Hydrographic Regions

The six hydrographic regions of the Niger Basin are distinguished by their unique topographic and drainage characteristics. Maps 2–7 in appendix 1 illustrate the hydrographic regions' geographic settings and characteristics. The regions are as follows:

- The Upper Niger River Basin and the Bani Watershed. The headwaters of the Niger have an extensive network of steep-sloped tributaries originating in Haute Guinée, whereas the Bani tributary network originates in the low-altitude plateaus of southern Mali and Côte d'Ivoire.
- The Niger River Inland Delta and Lakes District. This region is characterized by an immense, fertile, shallow-sloped alluvial floodplain with an extensive dendritic tributary network and shallow lakes.
- The Middle Niger, Malian-Nigerien, and Beninese-Nigerien Right-Bank Segment. This is a low-altitude plateau region with a series of tributaries that contribute to most of the Niger River's inflow along this segment.
- The Middle Niger Left-Bank Tributaries. This region is characterized by a wadi network in the upstream reach of this segment, with little contribution to the Niger River and an increased inflow from the tributary network in the lower reaches of the segment.
- The Benue River. This is a major tributary to the Lower Niger River originating in the high-altitude Adamawa Plateau in Cameroon.
- The Lower Niger River and the Niger Delta. Both these regions are located in a region of high rainfall, with an increase in the number of tributaries in the Lower Niger River, which flows south, emptying through the Niger Delta, an area characterized by swamps, lagoons, and navigable channels.

The Upper Basin of the Niger River and the Bani Watershed. The Upper Niger River Basin extends between latitudes 8°35' and 14° north and longitudes 4° and 11°30' west, with its downstream limit at Ké Macina and San (see details in appendix 1, map 2). Over the first 40 kilometers from its headwaters, the river drops 300 meters (an average slope of 7.5 meters per kilometer) toward the northeast. After passing the city of Faranah, it collects, in succession, the Balé and Tonboli tributaries, whose sources are located in the Fouta Djallon Massif. These tributaries are characterized by steep slopes, which explains the high flood levels of the Upper Niger. Upstream of Kouroussa, the Mafou is the first major tributary on the right bank of the Niger. It clears a series of rapids with a 10-meter drop along its downstream course. After passing the city of Kouroussa, it meets the Niandan, one of its main tributaries. Farther downstream, the Milo and the Tinkisso join the Niger from the right bank and the left bank, respectively, upstream of Siguiri. These are the largest Guinean tributaries of the Niger (although the Fie contributes more flow from Guinea, its confluence with the Niger is in Mali). Finally, the Sankarani joins the Niger from the right bank 40 kilometers upstream of Bamako, the Dion being its main tributary in Guinea.

At the border with Mali, the Niger River has already traversed 600 kilometers. With its tributaries and subtributaries, it drains more than 100,000 square kilometers of Haute Guinée, an area characterized by rough terrain at the headwaters (these highlands are particularly prone to erosion because human activity has largely reduced the vegetation cover). From Siguiri through Bamako to Koulikoro the flatness of the terrain, caused by long-term erosion, is broken in places by lateritic buttes in the southwest; plateaus cap a vast range of sandstone hills of the Mandingue region in the northwest, sloping toward the northwest with walls reaching heights of 300 and 400 meters above the alluvial plain. Most of these areas are poorly suited for agriculture; only the alluvial plain of the Niger valley allows agricultural development.

Between Bamako and Koulikoro, the Niger passes through two rapids (Sotuba and Kenie) before it flows gently into a vast plain that carries it to Segou and to the Markala Dam. Downstream of Segou-Markala, at Ké Macina, the river flows into the Inland Delta, with a riverbed slope of less than 2 centimeters per kilometer. At Ké Macina, the river has traversed about 1,000 kilometers from its source and has drained a watershed of 141,000 square kilometers.

The Bani, a primary tributary of the Niger, is created by the confluence of the Bagoé and the Baoulé, whose headwaters are located within Côte d'Ivoire. This area is characterized by very flat plateaus at relatively low altitude (between 280 and 500 meters, with 70 percent between 300 and 400 meters). The only hills of any significance in the area are west of the Bandiagara Plateau (reaching an altitude of 791 meters around Koutiala). The Bani's course is almost entirely located in southern Mali. Soon after the confluence, the Bani valley widens and the streambed almost disappears. The floodplain becomes about 10 kilometers wide downstream of Douna after the confluence with the Lotio-Banifing, which drains the Sikasso area. The Bani enters the Inland Delta just after San. The slope becomes extremely gentle, at less than 2 centimeters per kilometer. However, on the right bank the floodplain is limited by a series of sandstone hills from which several small streams flow. The Bani's course in the floodplain progresses toward the north where several secondary branches emerge, feeding the Djenné region. It finally empties into the Niger at Mopti, 1,300 kilometers from its headwaters. The Bani River Basin's area above Mopti is 130,000 square kilometers.

With nearly identical basin areas and dense hydrographic networks, the Bani and the Niger are distinguished by their respective slopes: The Niger has 24.4 percent of its entire surface area above 500 meters at Koulikoro, but the Bani has only 1.7 percent above this altitude and is characterized by much flatter topography.

The Niger River Inland Delta and Lakes District. The Inland Delta originates at Segou and ends in Koryoume, Timbuktu's river port. This area forms a huge parallelogram, located along a southwest-northeast axis, 400 kilometers long and 125 kilometers wide, averaging about 40,000 square kilometers. This area is without well-established watershed limits (see details in appendix 1, map 3). The Inland Delta can range from 30,000 square kilometers during periods of low water to 80,000 square kilometers during periods of high water. From Segou, for the Niger, and from Douna, for the Bani, the river network enters into an immense alluvial plain filled with various Quaternary and recent deposits. This area, known by different names (central delta, lake basin, inland basin, or Inland Delta of the Niger), is characterized by the alluvial deposits and multiple branches that are commonly found in deltas at the mouth of rivers. According to Gallais (1967), the Inland Delta is defined by the maximum extension of floodwaters and peripheral lakes:

- To the east and south by the slopes of the Bandiagara Plateau
- To the west by the dead delta, an area of ancient deposits above the current delta
- To the north by a series of dunes oriented east to west

The Inland Delta comprises four distinct morphologic regions—the upper delta, the central delta, the lakes district, and the lower delta. Other authors have described these regions in detail (Gallais 1967, 1979; Blanck and Lutz 1990; McCarthy 1993; Poncet 1994). In brief, the Inland Delta includes:

- The upper and central deltas, downstream from Ké Macina and Douna, combining two major branches. These branches of the Niger and Bani are sometimes referred to as the Malian Mesopotamia around Djenné in the upper delta, and Niger and Diaka for the plains of Kotia in the central delta. They extend to the lakes district (Lake Debo, Lake Wallado,

and Lake Korientze) and compose an immense discharge area largely inundated by the annual river flood.
- The lower delta that extends from the outlet of the lakes district with three main drainage points (Issa Ber, Barra Issa, and Koli Koli) all the way to Diré. Here a different geomorphology, characterized by superimposition over prior alluvial layers of a Holocene erg (sand sea) at Niafunke, with dunes oriented east to west, reveals a very diffuse water network. This network is dominated by interdune furrows that are often flooded and peripheral lakes fed by heavy floods.

The Inland Delta stores a volume of water varying from 70 cubic kilometers in wet years to 7 cubic kilometers in dry years. The flood arrives at Koulikoro in September and spreads slowly through the floodplain, and does not pass Diré until three months later. Water storage in the Inland Delta is accompanied by water loss (seepage and evaporation) of about 44 percent of inflows. The area of the peripheral lakes in the Inland Delta is particularly sensitive to the hydrological operation of the Niger River. These lakes were the subject of hydrologic studies during the 1950s, which was a period of very high water (Auvray 1960). However, the droughts during the 1970s and 1980s greatly reduced water availability, limiting refilling of the lakes, especially those on the right bank and Lake Faguibine (Guiguen 1985). The surface area of the lakes as reviewed in table 2.1 reflects the maximum extent, reached in the 1950s, a level that has not been reached again. By way of comparison, the surface areas of Lakes Debo, Wallado, and Korientze are 190, 70, and 57 square kilometers, respectively. The surface area of Lake Chad ranges between 5,000 and 25,000 square kilometers, 42 times the maximum surface area of Lake Faguibine and 11 times the total surface area of all the peripheral lakes of the Inland Delta of the Niger (Marieu, Kuper, and Mahieux 2000).

The Middle Niger, Malian-Nigerien, and Beninese-Nigerien Right-Bank Segment. As the Niger River exits the Inland Delta, it enters the Middle Niger River segment, which extends to Lokoja (refer to appendix 1, map 4). As it flows through the Niger Bend and after clearing the weir at Tossaye, the Niger River passes through the wadi network of the Vallée du Tilemsi and takes a southeast direction. On the Tossaye-Ansongo reach (212 kilometers), Quaternary alluvial deposits fill a valley that is 4 kilometers wide. On the left bank, windblown deposits obstruct the exits of wadis originating in the Adrar des Iforas, the Oued Essalaoua upstream of Bourem, and the Tilemsi upstream of Gao.

On the Ansongo-Niamey reach (352 kilometers), there is a succession of rapids at Fafa, Labezanga, and Ayorou, at 59, 109, and 144 kilometers from Ansongo, respectively, making this stretch one of the most difficult for navigation. The floodplain is about 2 kilometers wide. Labezanga marks the

Table 2.1 Geographic Location, Maximum Surface Area, and Capacity of the Peripheral Lakes of the Inland Delta

Lakes	Geographic location	Maximum surface area (km²)	Maximum capacity (M m³)
Tanda	15°45' N, 4°40' W	145	—
Kabara	15°45' N, 4°32' W	50	—
Tagadji	15°55' N, 4°08' W	90	—
Oro	16°12' N, 3°50' W	145	—
Fati	16°12' N, 3°42' W	140	—
Télé	16°27' N, 3°45' W	105	1,500
Faguibine[a]	16°47' N, 3°50' W	595	—
Gouber, Kamango	16°50' N, 3°40' W	120	—
Niangaye	15°50' N, 3°10' W	335	1,300
Do	15°55' N, 2°45' W	125	800
Garou[b]	16°03' N, 2°45' W	160	775
Haribongo	16°10' N, 2°45' W	55	290
Aougoundou	15°45' N, 3°18' W	85[c]	—
Korarou	15°20' N, 3°15' W	135	> 150

Source: Auvray 1960.

Note: — = not available.

a. The western limit of Lake Faguibine is set between the villages of M'Bouna and Tin Aicha. Since the droughts of the 1970s and 1980s, the water has never reached this limit. The villages of Raz el Ma and Adarmalane are therefore not included in this table.
b. The Garou Lake System includes Lakes Garou, Gakore, Tinguere, and Titolaouine.
c. During his visit on November 11, 1999, Marieu, Kuper, and Mahieux (2000) estimated the surface area of the lake to be 38 square kilometers. This value is not definitive because the lake was not completely full.

border between Mali and Niger. Downstream from Ayorou, the valley expands and collects, from its right bank, the first tributaries since the Bani. These include the Gorouol, draining the extreme north of Burkina Faso and the southern Mali region of Hombori, and two other tributaries originating in Burkina Faso—the Dargol and the Sirba.

On the Niamey-Malanville reach (336 kilometers), the first 100-kilometer section, to Kirtachi, crosses recent alluvial deposits. From Kirtachi to Boumba, the river traverses the Atakora sandstone massif for 100 kilometers through a narrow valley with abrupt changes in direction that form a "W" shape, from which the W International Park gets its name. From Boumba, the Niger flows through a wider valley (more than 4 kilometers wide) of alluvial deposits to the border with Nigeria. Before the W Park, the Niger collects three tributaries originating in Burkina Faso: the Goroubi, the Diamongou, and the Tapoa. After the W Park, the Niger, which now borders

Benin, collects the three main tributaries of its middle watercourse, draining the northeast of Benin: the Mekrou, the Alibori, and the Sota. On the left bank, endoreism is the most prevalent: Dallol Bosso (Azouâk), Dallol Foga, and Dallol Maouri are wadis, with no water reaching the Niger River.

The Middle Niger Left-Bank Tributaries. The Niger River continues in a southeast direction for 200 kilometers (see appendix 1, map 5) from the Niger-Nigeria border downstream to Yelwa. As it enters Nigeria, the Niger River collects some minor tributaries on the right bank (the Chodou, Wessa, and Kalia), all of which originate in Benin. A significant tributary on the left bank, the Sokoto, drains a large basin in the Sahelian region of Maradi, which encompasses the state of Sokoto-Rima and the northern portion of Kaduna state. Originating in the Gusau region, it collects the Rima—whose Nigerian section (Goulbins) in the upper basin starts in the dry valleys of the Aïr Massif—from the right bank, a short distance downstream of the city of Sokoto. From the left bank, the Sokoto-Rima River collects the Zamfara and the Ka.

After its confluence with the Sokoto-Rima River, the Niger River flows into a large plain, traveling north to south for 200 kilometers toward Jebba. Its course flows through the Kainji Reservoir for 130 kilometers. Between Kainji and Jebba, the Niger collects more minor tributaries on the right bank, the Oli and the Moschi, and various small watercourses on the left bank, whose contributions taken together are significant. In Jebba, a second dam cuts the flow of the Niger River. At this point, the Niger River has recovered the flow volume that it had when it left Guinea, over 2,700 kilometers upstream.

From Jebba to Lokoja, along a reach of nearly 400 kilometers, the Niger changes direction to the southeast and receives minor tributaries on the right bank—the Awun, Oshin, and Oro. About 150 kilometers downstream of Jebba, the Niger collects another major tributary from the left bank, the Kaduna, that drains an area of 65,500 square kilometers in the western part of the Jos Plateau. After taking a northwest direction, the Kaduna turns southwest, passing through the state capital of the same name, and collects several tributaries from the north, including the Mariga. With its steep slopes, the Kaduna is characterized by fast flows, strong floods, and severe lows that are indicative of a dry, tropical climate. At the end of this reach is Lokoja, where the Niger meets the Benue.

The Benue River. The Benue headwaters are at an altitude of approximately 1,300 meters on the north slope of the Adamawa Plateau in Cameroon (see appendix 1, map 6). The Benue collects the Mayo Rey and Godi streams from the right bank before crossing a vast plain of alluvial deposits, passing through the Lagdo Gorge, now the site of a large reservoir, then reaching a wide alluvial plain through which it flows into Nigeria.

Upstream of Garoua and 300 kilometers from its source, the Benue encounters the Mayo Kebi River that drains the marshy areas of the Kabia and the depression of the Toubouris Lakes in Chad, and channels strong torrents coming from the Mandara Mountains in Cameroon. The Mayo Kebi River is 420 kilometers long. At the confluence, the Benue takes a southwest direction, passes Garoua at an altitude of 175 meters and after flowing westward for a distance of 80 kilometers, it reaches the Nigerian-Cameroon border. There it collects the Faro, the powerful river from the Adamawa that is subject to strong floods and a large silt load. In Nigeria, the Benue passes Yola (60 kilometers inside the border); is supplied on the left bank by the Ini, a river coming from the Alantika and Shebshi Mountains; then collects (70 kilometers downstream from Yola) the Gongola, another strong tropical river coming from the northeast Jos Plateau, with a basin of 21,500 square kilometers. Like the Faro, the Gongola carries a heavy sediment load, making navigation difficult immediately downstream.

The Benue then takes a southwest direction, which it will follow for approximately 450 kilometers, to Makurdi. On this reach, the right bank tributaries coming from the Jos Plateau remain small, whereas significant rivers enter the Benue from the left bank. They come from the mountain chains and massifs between Nigeria and Cameroon (Shebshi, Gotel Mountains, and the Bamenda Massif; the highest point in the Niger River Basin is 3,008 meters at Pic Oku, north of Bamenda in Cameroon). These rivers become larger as they flow south; the areas receive heavy rains. After the Belwa and the Fan, the Benue collects, in succession, the Taraba, the Donga, and the Katsina Ala, whose watersheds are 21,500, 20,000, and 22,000 square kilometers, respectively. The upper Donga makes up the border between the northwest province of Cameroon and Nigeria, and the Katsina Ala and its tributary, the Metchum, contribute considerable flows. Between Garoua and Makurdi, the Benue's volume of flow increases eightfold. Over the 220 kilometers that separate Makurdi from the confluence with the Niger at Lokoja, the Benue collects short tributaries on the left bank and stronger tributaries such as the Mada and the Okwa, coming from the southern Jos Plateau, on the right bank. At Lokoja, the Benue meets the Niger; at this point, the two waterways are about the same size, although the Benue has traveled 1,200 kilometers and the Niger has traveled almost 3,800 kilometers from its source.

The Lower Niger River and the Niger Delta. At Lokoja, the Niger River enters the Lower Niger River segment, which includes the Niger Delta. From Lokoja, the Niger River takes a north to south direction for 200 kilometers; it receives only a few small tributaries, including the Anambra on the left bank, which drains a basin with significant rainfall. Onitsha is the last monitoring station on the river. The Lower Niger flows for another 100 kilometers and the lower valley progressively transforms into the vast Niger Delta covering approximately 30,000 square kilometers, with no fewer than 30 outlets to the ocean. The main course of the Niger takes the name of Nun as it crosses the Niger Delta and discharges to the Gulf of Guinea, 4,200 kilometers from its source in Guinea.

Navigable River Segments

The Niger's navigable waters are a strategic axis for commerce. Although commercial travel routes have changed over time, efforts to link the landlocked segments with canals and land routes have provided access to the river's resources to all. The main navigable segments of the river, from the Niger Delta upstream to the Upper Niger, are described below and illustrated in appendix 1, map 10.

- In Nigeria, the overall navigable network in the coastal lagoons and river branches of the Niger Delta is more than 6,000 kilometers; half of that network is part of the Niger and Benue network (Sanyu and others 1995). Along the length of the Niger, navigation is possible for large flat-bottomed boats upstream to Onitsha (1,127 kilometers from the ocean) throughout the entire year, and even farther upstream to Jebba (1,448 kilometers) from August to February. Lake Kainji is navigable for 130 kilometers. The Benue is navigable to Makurdi from June to December, and to Garoua in Cameroon from August to November.
- From Lake Kainji to the Niger border, the river is only navigable at very high water for 336 kilometers from the border of Nigeria (Malanville-Dolé) to Niamey from August to February. Upstream of Niamey, navigation is possible for another 170 kilometers from Tillabery to Meana. Rapids and rocks limit navigation over a 123-kilometer segment from Meana to Fafa, except in unusually high water. From Fafa to Ansongo, Gao and Tossaye (in Mali), medium-tonnage barges can follow a distance of 270 kilometers at high water, and navigation is generally permanent on the upper reaches for 327 kilometers between Tossaye and Timbuktu's river port of Koryoume.
- From Koryoume to Ké Macina, most of the main stem is navigable from September to December, although the Inland Delta is open to small fishing boats all year. In addition, several tributaries are navigable in high water, although when the water is low the channel in Lake Debo is not navigable. Within the Inland Delta, Mopti is the main port, on the Bani tributary. Large barge vessels can navigate over 200 kilometers upstream through the locks of the Markala Dam near Kirango from August to

January. From Koulikoro to Bamako, there is a continuously navigable waterway in high water (August to January).
- Upstream of Bamako, small fishing vessels can navigate seasonally (August to November) to Siguiri, and similarly, through Guinea, in the low valleys of the Niger River and the Milo and Tinkisso tributaries. There is no real commercial traffic established between Mali and Guinea.

There is great potential to develop the Niger Basin's transportation network to expand access to markets and commerce, and to increase labor flows. This potential is not fully exploited and existing navigational segments are underutilized. There is a need to improve the transportation network to foster regional integration among and between the countries.

Geology and Hydrogeology

Groundwater base flow heavily influences the Niger River, with dry season contributions primarily within the alluvial plains. In general, groundwater tables are affected by annual rainfall and soil permeability. Groundwater is extremely important for potable water supply (which is of excellent quality in most cases) for both urban and rural settlements. Several international and bilateral donors and nongovernmental organizations have financed research and development work on small aquifers and village water systems, although more needs to be done to understand the hydrodynamics of the aquifers in the Basin. The Basin's geology ranges from ancient Archean formations to recent alluvial deposits, each with its own hydrogeologic potential in the Basin's different hydrographic regions.

Upper Basin

An ancient geologic landscape of crystalline rocks characterizes the upstream area of the Basin and most of the river's right bank. Groundwater occurrence is very limited in these rocks, which are generally impermeable except where they have been fractured or are weathered, which creates small aquifer zones. Groundwater replenishment from the headwaters is therefore generally very low (Fontes and others 1991). Examples of these formations follow:

- An Archean base of granite, gneiss, and mica schist is found in the Guinean section of the Basin, northern Côte d'Ivoire, southwest Mali, most of Burkina Faso, and northern Benin, with a few basic intrusions (dolerite and gabbros) in Guinea at the Fomi site, and in Niger near Tillabery.

- Middle and Early Precambrian schist and quartzite appear in the lower valleys of the Niger tributaries in Guinea and Mali, in the tributaries of the Bani in Côte d'Ivoire and Burkina Faso, and southeast of the Niger Bend in Bourem, Gao, Ansongo, and in the Niamey Valley.
- Cambrian schist and sandstone extend from Bamako to Sikasso.
- Ordovician sandstone-quartzite and various sandstones are found in the Dogon region, on the Mandingue Plateau, and on all the plateaus between Koulikoro, Koutiala, and Bandiagara.

Inland Delta

Downstream of Koulikoro in the Inland Delta, north of Segou, and also in the Gondo depression east of the Dogon region, Quaternary and recent deposits mask the substratum and, in particular, the Eocene to Pliocene Continental Terminal. These recent deposits are either alluvial or dunelike Holocene ergs, with groundwater aquifers linked to the waterways. The Continental Terminal is a continuous stratum aquifer composed of claylike sandstone, sand, and clays, with good water quality. It appears on the left bank of the Niger at Goundam, Timbuktu, and Gourma Rharous, then continues through Bourem and Gao to Niamey and Gaya, with an extension north inclusive of the sedimentary basins of Taoudenni, the Azaouâd, the Tilemsi, and the Azaouâk. The Continental Terminal aquifer is the most significant aquifer in the Basin and is widely used, particularly in Niger. The stratum is immense: It is not unusual to see aquifer thicknesses of over 100 meters extending over tens of thousands of square kilometers. Underneath the Continental Terminal and the Eocene and Cretaceous layers of these sedimentary basins lies the Continental Shale Band aquifer, which borders the Niger River just north of Benin but is also present in the semiarid area of Mali and Niger. This aquifer has abundant groundwater, but is generally of poorer quality than that of the Continental Terminal.

Lower Niger

On entering Nigeria, the Niger flows through a series of sedimentary deposits: Tertiary deposits on the left bank—a continuation of the Continental Terminal observed in Niger, frequently with artesian aquifers—followed by Cretaceous deposits that continue to Onitsha on the Lower Niger. From Jebba, Quaternary alluvial deposits fill the large plains located on both sides of the river, extending in the Benue valley all the way to Chad. This sedimentary basin, with remnant marine properties from the Cretaceous era, rises toward the east throughout the Benue valley in Cameroon and Chad, the farthest reaches of the Cretaceous Sea. Schist, limestone, and sandstone make up the Cretaceous layer in these areas; deposits are thick and the aquifers are often of very good quality. From the Onitsha area, the Tertiary marine layer crosses through the Cretaceous layer, and is then covered by Quaternary sediments from the coastal plain and delta.

Outside this sedimentary basin, the African Precambrian base predominates in the Niger River Basin, in southwest Nigeria, and in the central area of the Jos Plateau, Abuja, Minna, Kaduna, and the upper Sokoto, and extending to the Adamawa and other mountainous boundary massifs in Cameroon. It is composed of gneiss, ancient granite, and quartzite, with limited groundwater potential except in weathered zones that can produce useful local supplies.

Soils

The three major soil types of the Niger River Basin, according to the French soil nomenclature (World Bank 1986), are ferralitic soils, tropical ferruginous soils, and hydromorphic soils (figure 2.1). The characteristics of these three types of soils determine the nature of agricultural productivity in the Basin. As a result, agricultural development varies in the Basin with the geographic distribution of the specific soil. The characteristics for each soil type are described below:

- Ferralitic soils are seen in the extreme west of the Guinean basin of the Niger, in the south of the Bani watershed, in the north of Benin, and in the major part of the Niger River Basin in Nigeria, including the Benue watershed. These are thick soils (from 3 to more than 10 meters thick), where geochemical changes are extensive and spread over many millions of years.
- Hardened lateritic layers can be seen on the surface or at short depths on top of a mix of the ferralitic and ferruginous soils; they are found in particular in Guinea and in southern Mali. This concreted and hardened horizontal layer results from an upward migration of ferrous oxides and from their precipitation. Limited spots of tropical brown soil or tropical black clay (vertisols) can also be found.
- Bleached layers of ferruginous tropical soils are seen in the north of the Bani watershed, on the periphery of the Inland Delta in Mali, in the east of Burkina Faso, and along the northern part of the Niger River Basin and Benue watershed in Nigeria and Cameroon. They are associated with ferralitic soils in the Upper Niger River Basin upstream of Bamako and in the Kaduna watershed in Nigeria. Alternating dry and wet seasons, characteristic of the watershed's climate, have caused discontinuous changes in the rock over time. The changing layers have a variable thickness but are always less than 3 meters deep.

Figure 2.1 Schematic Map of West African Soils

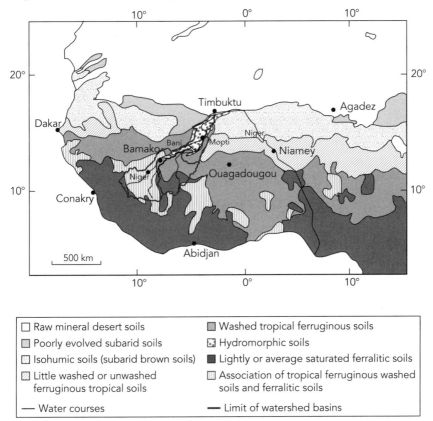

- ☐ Raw mineral desert soils
- ▨ Poorly evolved subarid soils
- ☐ Isohumic soils (subarid brown soils)
- ▨ Little washed or unwashed ferruginous tropical soils
- — Water courses
- ▨ Washed tropical ferruginous soils
- ▨ Hydromorphic soils
- ■ Lightly or average saturated ferralitic soils
- ▥ Association of tropical ferruginous washed soils and ferralitic soils
- — Limit of watershed basins

Source: World Bank 1986, modified by Picouet 1999.

The tropical ferruginous soils characteristic of a short wet season, which are little or not at all bleached, cover a large band in the north, from Mopti to Niamey and Maradi. Even farther north, subarid isohumic brown soils are present from Gourma to Gao (Mali), and in Niger. Sandy dunes, poorly evolved subarid soils, and lithosols characterize this section and indicate that the Niger River has reached the Sahara.

Hydromorphic soils, linked to the presence of a temporary or permanent aquifer that is close to the surface, can be found in lake basins, riverbeds, and low clay plains. Almost all of the soils in the Niger's Inland Delta are of this type; 74 percent are flooded every season in the active Inland Delta (PIRT 1983).

Natural Environment

The Niger River ecosystem corridor has provided people with sustained livelihoods even in the harshest drought conditions. The environment in its natural state and under pressure affects and is affected by the Niger.

The Natural Environment

Along its course, the Niger River traverses almost all the possible ecosystem zones in West Africa. The Niger River's headwaters are located at an altitude of 800 meters, at the fringes of the Guinean moist forests. The river then passes through woody savannas and areas of sedge vegetation. At the western edge of the Inland Delta is a short-grass savanna, with thorny shrubs and acacia wood, followed by a region of tussocky grass interspersed with dense wooded vegetation. The eastern Inland Delta is made up of Sudano-Sahelian flooded grasslands and a labyrinth of 50,000 to 80,000 square kilometers of wetlands and lakes. The floodplains are pastures of bourgou grass that support livestock, wildlife, and fish nurseries. In the wetlands, flora have adapted to extreme fluctuations in water levels. At the Niger Bend, the river reaches the fringes of the desert. The high rainforest belt begins farther south, at Onitsha. This belt merges below Aboh, in Nigeria, with mangrove forests and swamp vegetation in the Niger Delta. The Niger River system sustains an extensive biological community, hosting diverse ecosystems that harbor 36 families and nearly 243 species of freshwater fish, of which 20 are found nowhere else (11 of the 18 families of freshwater fish that are endemic to Africa are represented in the Niger River). Species in the river range from West African manatees and hippopotamuses to crocodiles. There is a rich collection of transmigratory birds found in the Inland Delta, among them black crowned cranes, herons, egrets, and storks; pelicans and flamingos are found in the upper Benue. The Niger Delta also includes an extensive mangrove forest.

The Natural Environment under Pressure

A combination of human population growth, unsustainable resource use and development, and desertification is threatening the Niger River's ecosystems and ability to supply crucially needed natural resources to the people of West Africa. River flow in the Basin is decreasing and fishing pressure is increasing, leading to drastic declines in production of fisheries. Deforestation and farming of fragile soils contribute through erosion to sedimentation in river channels. Waterborne diseases have increased and invasive aquatic species have spread, choking river channels.

Climatic Conditions

The Basin has two distinct seasons—a rainy summer and a dry winter—except for Nigeria, which has four seasons.[8] Situated between the equator and the Tropic of Cancer, the region is generally warm or hot, although the high mountains and Sahara Desert experience extreme temperatures. Along the coast, the annual average temperature range is 21°–28°C (70°–82°F); inland to the north the temperatures fluctuate more according to the season, with an annual average temperature range of 12°–29°C (54°–84°F). Given its geographic setting in West and Central Africa, the Niger Basin is characterized by the climatic conditions associated with the movement of air masses of the Intertropical Convergence Zone north and south of the equator. During the boreal summer (June to November), the rise of the Saint Helena high-pressure area toward the north signals the beginning of the monsoon season, with humid and unstable maritime equatorial air and relatively cool temperatures. The monsoons are longer and heavier in the southern part of the Basin. The boreal winter (December to May) is the dry season; under the influence of a Saharan high-pressure zone, the northeastward harmattan wind brings hot, dry air and high temperatures, which last longer in the north. Annual rainfall ranges from fewer than 100 millimeters (4 inches) in the Sahel zone to more than 1,200 millimeters (48 inches) along the pure tropical areas in the Guinea zone. The regional climate classification as defined by annual rainfall, according to Maley (1982), and reviewed and modified by L'Hôte and Mahé (1996), is noted in table 2.2.

The upper half of the Niger River Basin encompasses five climate zones (see figure 2.2), which are based on the duration and abundance of annual

Table 2.2 Climate Classification for West Africa from South to North

Annual rainfall (mm)	Climate classification for Western Africa	Climate classification	French classification[a]
>1,200	Sudanese II and III	Transitional tropical	Tropical de transition
750–1,200	Sudanese I	Pure tropical	Tropical pur
300–750	Semiarid south	Semiarid tropical	Tropical semi-aride
150–300	Semiarid north	Semiarid desert	Semi-aride désertique
<100–150	Saharan	Desert (arid)	Désertique

Sources: Maley 1982; L'Hôte and Mahé 1996.
a. See figure 2.2.

Figure 2.2 Geographical Distribution of Different Types of Climate in Africa

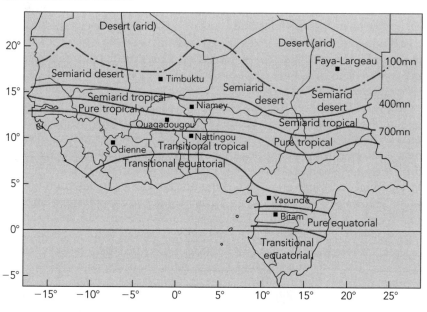

Source: L'Hôte and Mahé 1996.

rainfall. The rainy season is centered in the month of August for all these climate zones. The five zones as they extend across the region can be described as follows:

- The Guinean region, which includes the headwaters of the Niger River Basin and its tributaries, is characterized by a transitional tropical climate (*tropical de transition*) that is often called the Guinean Climate, where the annual rainfall is greater than 1,200 millimeters.
- The region encompassing Siguiri-Sikasso is characterized by a pure tropical (*tropical pur*) climate, with annual rainfall between 750 and 1,200 millimeters.
- The areas around Mopti are characterized by a semiarid (*semi-aride*) tropical climate, with annual rainfall between 300 and 750 millimeters.
- The areas of the Inland Delta around Timbuktu have a semiarid desert climate (*semi-aride désertique*), with annual rainfall between 150 and 300 millimeters distributed over three to four months.
- A part of the Inland Delta is characterized by a desert climate (*désertique*), with less than 150 millimeters of annual rainfall distributed over the three summer months.

After the Niger Bend, the river crosses all of these climate zones in reverse, with the same type of annual rainfall distribution:

- The desert region, including the landlocked wadi valleys in Mali and Niger, is limited in the south by the Malian-Nigerien border.
- The semiarid desert, which includes Burkina Faso, continues all the way to Benin and northern Sokoto in Nigeria.
- The semiarid tropical region covers northern Benin and the latitudinal band extending from Gaya (Niger)–Kainji (Nigeria) to Maroua-Garoua (Cameroon).
- The pure tropical region covers most of the Basin in Nigeria, the Jos Plateau, and the Adamawa.
- The transitional tropical region covers lower Nigeria.

Appendix 1, map 11, illustrates the average annual rainfall patterns in the Basin and appendix 1, map 12, illustrates representative annual rainfall distribution throughout the Basin for both dry and wet months. Additional regional and site-specific information on climatic characteristics, annual and monthly rainfall, temperature, humidity, and evaporation rates in the Niger River Basin can be found in appendix 2.

Climate and Water Resources Variability

The variable duration of the rainy season and significant rainfall deficits in this large basin result in hydrological deficits that are not necessarily reflected in a direct response of the base flow.

Variability in Rainfall

Studies on climate variability in West Africa show a significant decrease in both the amount of annual rainfall and the duration of the rainy season. Carbonnel and Hubert (1992) detect a 19-year climate variation for the period 1970–89. L'Hôte and Mahé (1996) compare the rainfall average during the period 1951–69 with the period 1970–89 and determine a southward movement of isohyets in the range of 150–250 kilometers, depending on the Basin climate zone. Figure 2.3 highlights this southward movement of rainfall variability. Furthermore, analysis of monthly rainfall data for the whole region by Le Barbe and Lebel (1997) shows that the dry period is characterized by a decrease in the number of rainy events, but the mean storm rainfall varies little. A deficit of 10 percent to 30 percent in rainfall generally leads to a deficit of 20 percent to 60 percent in river discharges, confirming that rainfall in the Basin varies considerably but river discharges vary still more.

Figure 2.3 Highlights of Isohyets Moving Southward

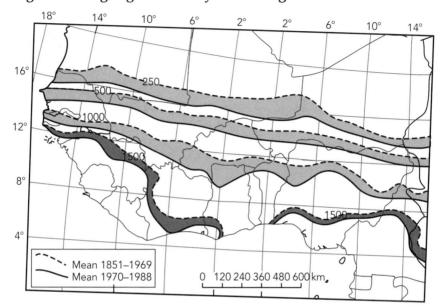

Source: L'Hôte and Mahé 1996.

Relationship between Rainfall and Runoff

Olivry (1998) notes that the long-term relationship between rainfall and river flow is largely influenced by groundwater base flow, as is the case of the Niger River. Cumulative dry periods contribute to a reduction in base flow, and a return to sustained river flow requires replenishment of the aquifers, which is possible only with cumulative rainy years. Regionally, this return is referred to as the "memory of the river." Several studies in the Niger and Bani upper basins confirm the correlation between decreased rainfall and low river flows, as illustrated in figure 2.4. Because groundwater base flow varies with, and responds to, the rainfall from previous years, the river flow in turn fluctuates with the level of aquifers, especially during a series of dry years. The dry years in the early 1970s, known in western Africa as *la grande sécheresse* (the great drought), saw the flow of the Niger River decline to unprecedented low levels. Yet a subsequent decrease in the rainfall deficit, or increase in rainfall, in the latter half of the 1980s does not correlate with the flow variation (figure 2.5). The Niger River's delayed hydrological responses illustrate that it takes more than a good rainy year to return the river to its previous flow.

Figure 2.4 Correlation between Annual Rainfall and Runoff on the Niger River at Koulikoro

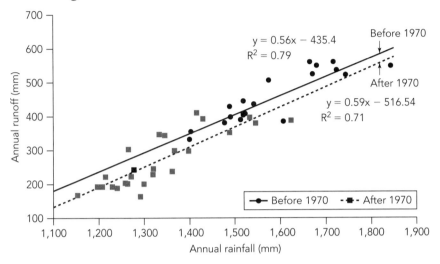

Source: Olivry 1997.

Note: y = regression; R^2 = squared correlation.

Figure 2.5 Yearly Variation of Average Rain and Flow Indices for Sudano-Sahelian Africa since the Beginning of the 20th Century

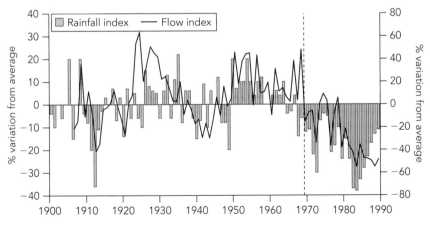

Source: Olivry and others 1993.

3
The Niger River Basin's Water Resources

Hydrology

From available information, chapter 3 evaluates the Niger River Basin's water resources in the six hydrologic reaches. This chapter examines the hydrology and water quality of the Niger River in some detail, using the same hydrographic regions described in chapter 2.

Introduction

The Niger's hydrology is as extraordinary as its hydrography, undergoing remarkable changes in its hydrologic characteristics as it travels from its headwaters to the Gulf of Guinea. From a simple tropical system with abundant rainfall at its headwaters in the Upper Basin, the Niger moves northeast into the Inland Delta, losing both flow volume and velocity as it meanders near the Sahara. Then, as it flows southeast after the Niger Bend, inputs from other tributaries make up these losses little by little. After its confluence with the Benue, the Niger River continues as a large river to the Niger Delta. Appendix 1, map 13 illustrates the flows at various points along the Niger River Basin, including the significant losses within the Inland Delta and around the Niger Bend. Unlike other major West African rivers, such as the Senegal and the Volta, the lower reaches of the Niger River are sustained during periods of low flow in the spring by the arrival of the waters from the previous summer's flood in the upper Basin. This phenomenon is known as the black flood in Niger. The annual flood in Nigeria, rich in sediment, is in phase with the summer rains and is known as the white flood. This process is illustrated in figure 3.1 and described by Pardé (1933; translation of this passage by the authors):

> The lower Niger, by its curve at two ends, leads one to believe in double supply, and this is a false impression. In reality, throughout its basin, this river has only one rainy period, that of a tropical summer. But the particularities of its topography and profile, in length and breadth, curiously doubles the high water season. The Senegalese Niger near Bamako and the lower stretch, downstream of Say, experience a prominent summer flood

Figure 3.1 Example of the Hydrograph of the Niger River at Its Entry to the Niger Delta

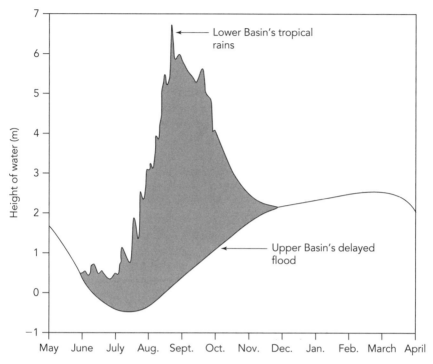

Source: Pardé 1933.

cresting in September. But the volume set in motion along the upper water course soon slows significantly, and partially dries up through evaporation and infiltration, before the river's "Big Bend" at Timbuktu, because of the gentle slope and the enormity of the flood plain where the water disperses and almost comes to a halt. The main flow, which passes Koulikoro generally about September 25, does not arrive in Timbuktu until about January 1; then it does not reach Niamey, when it is very low, until about February 2, six months after the rains that produced it. On the lower Niger, this Senegalese flood, slowed down in a manner not seen anywhere else in the world, does not participate in the local summer flooding; but it slows the drop in water, from November or December; it ends up by causing a gradual rise up to a maximum in March that is significantly lower than that of September. This is the most impressive example of the influence of topography on the flow conditions of a water system.

For the purposes of description, the Niger River is divided into six hydrologic regimes, to match the six hydrographic regions described in chapter 2.

- The Upper Niger River Basin and the Bani Watershed
- The Niger River Inland Delta and Lakes District
- The Middle Niger, Malian-Nigerien, and Beninese-Nigerien Right-Bank Segment
- The Middle Niger Left-Bank Tributaries
- The Benue River
- The Lower Niger River and the Niger Delta

Appendix 1, map 14 shows the location of the Hydrological Forecasting System in the Niger River Basin (HYDRONIGER) data collection and monitoring stations along the river and major tributaries. Appendix 1 also includes maps with geographic details of these regions (maps 2–8), and appendix 2 provides detailed technical supporting data and information on flows, rainfall, evaporation, and sediment transport in the Basin. Appendix 3 provides an overview of the history of data collection and management in the Basin.

The Upper Niger Basin

The Upper Basin of the Niger River and the Bani Watershed (see appendix 1, map 2) contains four primary tributaries of comparable size: the Niger, referred to as the Djoliba (watershed of 18,600 square kilometers); the Niandan (12,700 square kilometers); the Milo (13,500 square kilometers); and the Tinkisso (19,800 square kilometers). The first three watersheds receive abundant rain, sometimes more than 2,000 millimeters per year at the headwaters; they are also steep, with high levels of mean annual runoff: 563 millimeters on the Milo, 531 millimeters on the Niandan, and 442 millimeters on the Niger at Kouroussa. Mean annual runoff for the Tinkisso at Ouaranin in the north, by contrast, drops to 244 millimeters. These values were calculated over the period of 1950–2000 and include both wet and dry periods (Rodier 1964; Bamba and others 1996; Sangaré 2001).

The Upper Basin above Siguiri has a surface area of 67,600 square kilometers and a mean annual flow of 948 cubic meters per second, equaling an annual runoff of 438 millimeters. Over the same 50-year period (1950–2000), the rain received by the watershed was 1,520 millimeters per year; the flow deficit is 1,082 millimeters, which can be attributed solely to actual evapotranspiration. The transitional Guinean Climate system explains the sustained flows observed from June to January, with several flood peaks and a maximum flow typically September–October; the low-flow season lasts only four months, with the lowest level April–May. Table 3.1 gives the mean monthly flows of the four main branches of the Niger in Guinea.

The Banankoro station monitors the flow of the Niger River entering Mali, which differs only a little from the flow at Siguiri. Farther downstream, the Niger receives the Sankarani River. The Sankarani watershed, two-thirds

Table 3.1 Mean Monthly Flows at the Four Main Branches of the Niger River in Guinea

cubic meters per second

Station	Jan.	Feb.	March	April	May	June	July	August	Sept.	Oct.	Nov.	Dec.	Avg. year
Niger to Kouroussa	52.6	25.4	13.5	8.76	12.3	57.4	180	414	682	595	299	112	232
Niandan to Baro	49.1	26.3	17.0	14.6	30.5	108	257	464	679	544	282	103	215
Milo to Kankan	35.2	20.0	14.4	16.0	29.2	81.6	229	439	599	412	179	70	177
Niger to Siguiri	200	103	58.6	44.0	67.1	231	804	2,054	3,304	2,708	1,244	454	948

Source: Brunet–Moret and others 1986.

of which are in Guinea, covers 35,500 square kilometers at the confluence; the river is monitored at the Selingue gauging station.

The first monitoring station on the Niger River was installed at Koulikoro in 1907, providing a valuable historical reference. The record of the Niger's flow at this station and the record of the flow at the Bakel monitoring station on the Senegal River together make up the longest and most complete dataset that exists, showing hydroclimatic variations in West Africa since the beginning of the 20th century. Above the Koulikoro gauging station, the watershed has a surface area of 120,000 square kilometers, one-fifth of which is in Mali. The mean annual flow calculated over 83 years is 1,420 cubic meters per second, providing a specific runoff flow (calculated as surface area) of 11.8 liters per second per square kilometer. The annual runoff of 370 millimeters is 25 percent of the mean annual rainfall, estimated at 1,500 millimeters, with losses caused by evapotranspiration of 1,130 millimeters (Brunet-Moret and others 1986). The largest floods were observed in 1924 and 1925 at 9,409 and 9,669 cubic meters per second, with the last major flood (at 9,344 cubic meters per second) in 1967. The mean annual low flow over 73 years, before the Selingue Dam, was 25 cubic meters per second.

The Bani Basin

The Niger River is joined by the Bani River, an important tributary, at the Inland Delta. The Bani is monitored at the Douna station (watershed of 101,600 square kilometers). Between 1953 and 1990, the mean annual flow was 419 cubic meters per second or a specific runoff flow of 4.12 liters per second per square kilometer, three times lower than that of the Upper Niger. The mean annual runoff was 130 millimeters, or 10.8 percent of mean rainfall of 1,200 millimeters. Although having a similar area to the Upper Niger above Koulikoro, the Bani watershed receives less rain and has much lower runoff. The Bani flow therefore represents only 11 percent to 41 percent of the total flow at Koulikoro, depending on the year.

Seasonal Variability

The rainfall pattern over the Upper Niger and Bani watersheds creates a large seasonal variation in flows and monthly distribution of runoff in the watersheds, and causes significant variations between low-water and flood stages. Along the Niger, for six months (January through June) low flows represent less than 8 percent of the total annual flow. The increase in flow begins in May but does not become significant until July. The monthly runoff coefficients (that is, runoff as a percentage of rainfall) are 17 percent, 30 percent, and 25 percent in August, September, and October, respectively. The highest flood level generally occurs during the second half of September, a slight delay from the maximum rainfall in August. More than 80 percent of the annual flow is accounted for during August to November. The flood recession, rapid and rather regular, is characterized by two phases. The first phase corresponds to the depletion of surface water; the second phase, which is characterized by a rapid drop in base flow at the end of November, corresponds to the seasonal depletion of shallow aquifers.

This cumulative draining of small shallow aquifers, whose recharge depends solely on infiltration of rainwater runoff, is characteristic of the general geomorphology of intertropical Africa (Olivry, Bricquet, and Mahé 1998). A study covering nine representative watersheds in Mali shows that, for some, an important part of the flow originates from delayed base flow, with inflows from draining of marshes and from groundwater storage, notably in southern Mali and northern Côte d'Ivoire (Joignerez and Guiguen 1992).

Flooding in the Upper Basin

Both the Upper Niger River and the Bani River are subject to large annual floods (Rodier 1964). Heavy rainfall, which is spatially limited, does not necessarily correspond to high annual flows, due to the large size of the watersheds. The annual maximum flow does correlate well with the annual runoff (Olivry, Bricquet, and Mahé 1998). The frequency analysis for average flows and highest flood flows in wet, average, and dry years is presented in table 3.2.

Floods in the Upper Basin have been as high as 1,500 cubic meters per second on the Tinkisso at Ouran, 1,000 cubic meters per second on the Milo at Kankan, and 1,500 cubic meters per second (and 1,960 cubic meters per

Table 3.2 Frequency Analysis of Observed Hydrological Parameters in the Upper Niger River Basin
cubic meters per second

Recurrence interval	Wet years			Average years	Dry years		
	100	20	10	2	10	20	100
Average flows							
Koulikoro (Niger River)	2,366	2,089	1,940	1,419	898	750	472
Douna (Bani River)	918	854	827	419	153	84	70
Highest flood							
Koulikoro (Niger River)	9,330	8,290	7,735	5,590	3,800	3,300	2,260
Douna (Bani River)	4,460	3,560	3,480	2,425	806	565	364

Source: Olivry and others 1995.

second in 1962) on the Niandan at Baro. The high flood levels on these tributaries generally occur in the latter half of September and are almost in phase with the annual flood arriving at Siguiri, where the highest recorded flow level reached 5,930 cubic meters per second in 1962, with the 10 highest recordings exceeding 5,000 cubic meters per second over the past 50 years. The lowest flow levels, which had been, on average, 50–60 cubic meters per second for the Niger at Siguiri, fell to fewer than 20 cubic meters per second in the last quarter of the 20th century.

Despite the occasional high flows, the generally low-impact "flooding power" of the Upper Niger and Bani is typical of watercourses in arid Africa. These two watercourses drain different areas in terms of topography and rainfall; thus, their floods do not have the same power, nor do they occur at the same time. However, the upper basins of both the Niger and Bani can generally be characterized by gentle slopes, low permeability, floodplains that absorb overflows, and the episodic nature of the rainy season, referred to as the African monsoon (Rodier 1964). The Upper Niger system is heavily dominated by the rainfall in the Upper Guinean Basin, and lower rainfall in the Bani watershed explains its lesser contribution.

The Inland Delta and Lakes District

The Inland Delta, with its system of lakes on both banks of the Niger River, is the result of the immense discharge from the Upper Niger Basin and Bani tributary. The Inland Delta covers approximately 40,000 square kilometers—of which 20,000–30,000 square kilometers are floodplains—but can expand up to 80,000 square kilometers (see appendix 1, map 3). The hydrological characteristics of the Niger River's Inland Delta and lakes district are largely dependent on

- Exogenous runoff conditions, with most of the water resources coming from upstream areas with higher rainfall; and
- Morphological and climatological conditions specific to the Inland Delta, affecting runoff (water loss, flooding) and water balance (evaporation, infiltration).

A comparison of the average annual flows in three typical years (high, average, and low) from Koulikoro to Tossaye is summarized in table 3.3. The year 1954 corresponds to a typical wet year, 1968 to an average year, and 1985 to a typical dry year. An assessment of these flows shows that runoff, monitored at the entry of Diaka and after the Bani confluence at Mopti, had already lost about 18 percent (during a wet year), 14 percent (during an average year), and 6 percent (during a dry year) of its initial contribution. The losses were even more significant when the flooded area

Table 3.3 Example of Flow Progression from Koulikoro to Tossaye for Three Different Years

cubic meters per second

Station	Wet year, high flow (1954)	Average year, average flow (1968)	Dry year, low flow (1985)
Koulikoro	2,075	1,445	915
Ké Macina	1,951	1,306	765
Bani Douna	926	456	150
Bani Sofara	646	382	130
Diaka/Kara	642	409	255
Niger Mopti	1,702	1,098	604
Diré	1,522	1,118	619
Tossaye	1,457	1,033	574

Source: Olivry and others 1998.

increased, with greater inflows from secondary tributaries. In relation to input from the Upper Niger at Ké Macina and the Bani at Douna, the flows at Diré, at the downstream end of the Inland Delta, show a loss within the delta of about 47 percent (a wet year), 37 percent (an average year), and 32 percent (a dry year).

Evaporation in the Inland Delta

The Inland Delta has an estimated storage capacity varying from 7 cubic kilometers to 70 cubic kilometers, with a high rate of loss caused by evaporation over the thousands of square kilometers of its floodplains. This loss, estimated at about 44 percent of the inflow, constitutes an important source of evaporation in West Africa. The regression in figure 3.2 illustrates the correlation between flow input to the Inland Delta (at Ké Macina and Douna) and flow at the outlet (at Diré), thus reflecting the water losses in the Inland Delta (see appendix 1, map 13).

An assessment of annual volume losses shows that they can reach 25 cubic kilometers between entry into the Inland Delta and the outlet at Diré for a wet period, and 7 cubic kilometers for a dry period, corresponding to a ratio of 14 to 4. The range of extremes has been from 40 cubic kilometers to 6 cubic kilometers in annual volume loss. For example, during a 25-year dry period (1970–95), the base flow of the tributaries dropped 46 percent against a drop in rainfall of only 19 percent. Outflows represented 54 percent of inputs in wet periods and 65 percent in dry periods. In other words, losses in the Inland Delta are lower in absolute value but also in relative value in dry years.

Figure 3.2 Correlation between the Input at Ké Macina/Douna and the Losses at Diré in the Inland Delta from 1955 to 1997

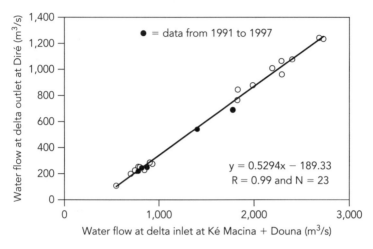

Source: Picouet 1999.

Note: y (regression); R (squared correlation); N (rank).

Figure 3.3 Evolution of the Yearly Volume Loss in the Inland Delta from 1955 to 1990

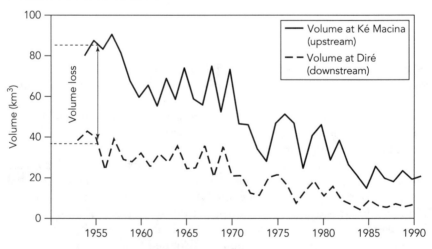

Source: Olivry and others 1995.

Note: This volume loss corresponds to the difference between the upstream volume (at Ké Macina) and the corresponding downstream volume (at Diré).

Table 3.4 Average Values of Rainfall and Evaporation in the Inland Delta for Wet Periods and Dry Periods
millimeters

	May	June	July	Aug.	Sept.	Oct.	Nov.	Dec.	Jan.	Feb.	March	April	Year
Pw	17	58	94	190	92	26	0	3	1	0	4	6	490
Pd	13	50	92	97	65	8	2	0	0	0	0	3	330
Ew	220	210	200	160	165	185	180	160	165	185	210	220	2,260
Ed	240	220	210	180	170	195	180	160	170	190	215	230	2,360

Source: Olivry and others 1995.

Note: P (rainfall); E (evaporation); w (wet periods); d (dry periods).

Figure 3.3 illustrates the difference between the annual volume in the Inland Delta from Ké Macina to the outlet at Diré. In an average year, 60 cubic kilometers arrive from upstream and 30 cubic kilometers are lost by evapotranspiration, corresponding to the extent of the flooded area in this reach, which can vary from 5,000 square kilometers to more than 30,000 square kilometers from the driest to the wettest years. Using information from table 3.4, figure 3.4 confirms the significance of water losses in the Inland

Figure 3.4 A Comparison of Mean Monthly Volume Losses in the Inland Delta for Representative Dry and Wet Years

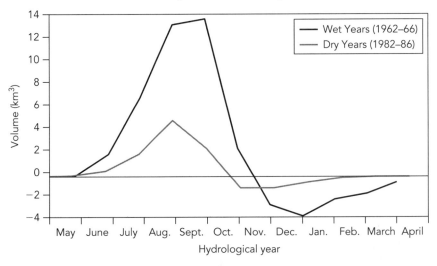

Source: Olivry and others 1995.

Note: From December, negative values indicate the restoration to the system from some of the waters stored in the floodplains.

Figure 3.5 A Comparison of Two Hydrographs on Inflow (Ké Macina+Douna) and Outflow (Diré) for Two Contrasting Years (1992 to 1993 and 1994 to 1995)

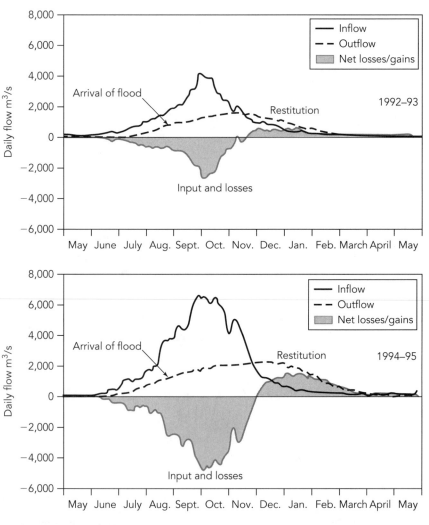

Source: Olivry and others 1995.

Delta by contrasting the mean monthly volume losses for representative wet and dry years. This is further substantiated in the hydrographs of inputs and outflows from the Inland Delta for two contrasting years (figure 3.5). The changes in surface area in the Inland Delta have been confirmed by satellite imagery on the United States National Oceanic and Atmospheric Administration's Advanced Very High Resolution Radiometer (NOAA-AVHRR) (see appendix 1, map 15; Mariko and others 2000). Table 3.4 presents the amount of rainfall and evaporation calculated for the Inland Delta in wet and dry periods, based on climate data from Mopti, Timbuktu, Diré, and Niafunke.

Flooding in the Inland Delta

Another characteristic of Inland Delta hydrology is the cushion it provides during the annual floods by slowing the pace of the flow, which spreads out in both space and time. The larger the flood, the more the flow spreads out over the space of the floodplain, and the longer it takes to spread out over time, with maximum decreases in flow downstream appearing later in the season, as illustrated for several monitoring sites for the entire Inland Delta (figures 3.6 and 3.7). In general, the peak flow period that arrives in September is delayed as it spreads, exiting the Delta three months later. A phase of receding water extends into February. The downstream impact of flooding on the Middle Niger is such that during a dry year the maximum flow arrives in Niamey in mid-December, whereas in a wet year the maximum flow is not seen until the end of January or early February.

Figure 3.6 Hydrographs Illustrating Buffering from Upstream to Downstream for the 1992 to 1993 Hydrologic Year

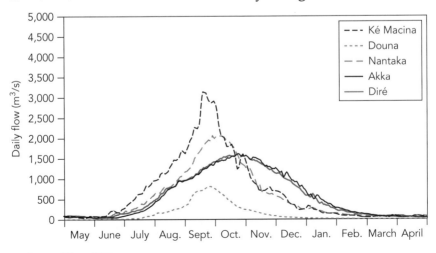

Source: Olivry 2002.

Figure 3.7 Hydrographs Illustrating Buffering from Upstream to Downstream for the 1994 to 1995 Hydrologic Year

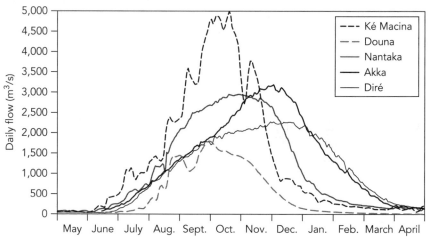

Source: Olivry 2002.

The Middle Niger

Downstream from the Inland Delta is the beginning of the middle reach (Malian-Nigerien and Beninese-Nigerien, and its right-bank tributaries), where the Niger changes course toward the southeast at Tossaye-Bourem without receiving any additional inflow. At Niamey, the maximum flows are usually twofold: a first wet seasonal peak flow and an upstream dry seasonal peak flow. The first high-water discharge, known as the white flood, occurs soon after the local rainy season in September; a second rise—the black flood begins in December—with the arrival of the delayed flood from upstream. May and June are the low-water months in the Middle Niger.

The first tributary in this reach, the Gorouol, comes from Burkina Faso. The Gorouol contributes, in an average year, 0.1 cubic kilometer for a watershed of 45,000 square kilometers (0.07 liter per second per square kilometer, or 22 millimeters of runoff), which is equivalent to a day's worth of evaporation in the Inland Delta. Before approaching Niamey, there are three tributaries from Burkina Faso that contribute about 1 cubic kilometer in an average year. In this middle reach, the river system changes very little, except in September when the semiarid tributaries flood. The first flood peak reinforces the Niger River's flow by 5 percent to 20 percent. Historically,

Figure 3.8 Hydrographs for Eight Years (1994 to 2002) at Niamey Illustrating the Two Flooding Phases in the Middle Niger

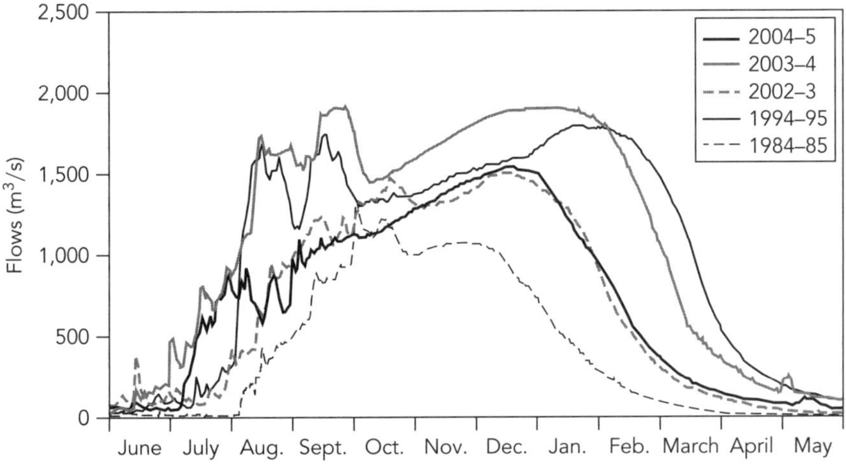

Source: WHYCOS-WMO/CIP-NBA (database).

the average flow at Niamey was 1,020 cubic meters per second, but over the past 20 years it has been only 670 cubic meters per second, or two-thirds of the previous average. The maximum annual average was 1,840 cubic meters per second. Frequently, the floods caused by the input from the semi-arid tributaries in September are higher than the delayed flood from upstream. Low-water flows have persisted since the 1970s, and the Niger River even ceased flowing in 1985. Figure 3.8 shows annual (1994–2002) hydrographs at Niamey, beginning with low flows in July, for which the first maximum floods correspond to the nearby Burkina Faso tributaries, whereas the second maximum corresponds to the delayed flood of the Upper Niger from December to February.

From Niamey to the border with Nigeria, the river gains 20 percent of its flow from right-bank tributaries coming from Benin: the Mekrou, the Alibori, and the Sota. The total annual runoff for these watercourses is on average 100 millimeters, but with significant interannual irregularity: 10-year mean annual runoff figures for dry and wet periods range from 40 millimeters to more than 200 millimeters (Le Barbe and others 1990). At Malanville, the annual flow volume was an average of 36 cubic kilometers, or 1,140 cubic meters per second (over 35 years), although the interannual flow has dropped to 800 cubic meters per second since the 1980s. The floodwaters from the Benin tributaries arrive in July, bringing sustained low-water flows greater than the flows that arrive from Niamey. The black floods and white floods are of equal average values (2,200 cubic meters per second). The white flood 10-year frequency is estimated at 2,800 cubic meters per second.

The Nigerian Stretch of the Middle Niger and Its Left-Bank Tributaries

In Nigeria, the Niger River continues to grow with the contributions from its rainfed tributaries. Downstream from the Sokoto on the left bank, which drains a partially semiarid, partially tropical, basin upstream of the Kainji Dam, the average annual flow is significantly increased by other more southern contributors. At Jebba, the Niger has a long-term mean annual flow of 1,600 cubic meters per second, although this has dropped to an average of 950 cubic meters per second over the past 15 years.

After receiving the Kaduna (watershed of 65,500 square kilometers) with a flow of 600 cubic meters per second (and floods of 3,000 cubic meters per second, on average), the Niger flows to Baro. Over the period 1914–60, the average flow was 2,500 cubic meters per second, or a flow of 79 cubic kilometers for a 730,000-square-kilometer watershed at Baro. The annual maximum flow reaches 9,000 cubic meters per second and the 10-year flood flow reaches 12,000 cubic meters per second. As at Koulikoro, the hydrograph peaks with a maximum in September for the white flood, whereas the black flood is muted but still noticeable because the low flows are sustained from January to May at a level of 1,500–2,000 cubic meters per second.

The Benue Basin

At Lokoja, the Benue River reinforces the Niger (appendix 1, map 6). On the Benue, there is only one high-water season. Because of the Benue's more southerly climatic location, this normally occurs from May to October—earlier than on the Middle Niger. Over the period 1950–80, the Benue had an average flow at Garoua (Cameroon) of 350 cubic meters per second, of which 250 cubic meters per second was from the Lagdo Dam, measured at Riao, and 100 cubic meters per second from the Mayo Kebi (Northern Cameroon and Chad). Using intermittent data obtained since 1980, the average flow at Garoua would be 330 cubic meters per second or an annual flow of 10.4 cubic kilometers. In normal conditions, floods could reach extraordinary maxima (6,000 cubic meters per second in 1948) at the end of August or September, but the median maximum is approximately 2,900 cubic meters per second. Low flow could be down to only tens of liters per second, in other words, almost no flow. Coming from Cameroon, the Benue receives inflow from the Faro tributary just before the Nigerian border (mean annual flow about 310 cubic meters per second), and contributes 22 cubic kilometers in an average year (of which only 1.6 cubic kilometers comes from Chad). The total runoff is 230 millimeters for a rainfall of 1,240 millimeters, giving a runoff coefficient of 18.6 percent (Olivry 1986).

In Nigeria, there is an extensive network of tributaries that flow into the Niger River. First, on the right bank, comes the Gongola, with a mean annual flow of 200 cubic meters per second (maximum average flood approximately 1,200 cubic meters per second), followed by the left-bank contributions that come from much wetter mountainous areas (Cameroon ridges, Adamawa, and Jos Plateaus on the right bank) subject to a more southern transitional tropical climate. These tributaries—the Taraba, the Donga, and the Katsina Ala—have a total flow contribution of about 1,700 cubic meters per second in an average year, or 54 cubic kilometers for 63,500 square kilometers of basin surface area, giving a total annual runoff of 844 millimeters. Of this total, 14 cubic kilometers come from the northwestern section of Cameroon where the Niger River headwaters cover only 8,750 square kilometers of surface area, giving a mean annual runoff of about 1,600 millimeters from a mean rainfall of 2,600 millimeters (a very high runoff coefficient of 60 percent). Measured near their confluence with the Benue, the average flood flows are 1,800 cubic meters per second for the Taraba and the Donga, and 2,800 cubic meters per second for the Katsina Ala.

At Makurdi, the Benue River is transformed, reaching a mean flow of 3,150 cubic meters per second (100 cubic kilometers per year) for a 305,000-square-kilometer watershed. Over the past 20 years, this annual flow has been maintained (at 97 cubic kilometers), indicating that the incidence of drought has been lower here than elsewhere in West Africa (except during the 1980s). The average absolute low flow is 240 cubic meters per second and the average annual flood flow reaches 12,000 cubic meters per second. At the confluence with the Niger, over the same period, the Benue has a mean annual flow of 3,400 cubic meters per second (107 cubic kilometers).

The Lower Niger River and the Niger Delta

After Lokoja, the Lower Niger River—with an average annual volume of 190 cubic kilometers—flows directly south toward the Niger Delta and the Gulf of Guinea. The Lower Niger has a high-water period that begins in May or June (caused by high rainfall in the Benue basin)—and a low-water period that is at least a month shorter, because the rains in the south start earlier. At the last monitoring station, Onitsha, river flows have increased to a total of 200 cubic kilometers per year. The compilation of data from Baro and Makurdi shows a progressive increase in flow in June–July, with maximum levels reached in October (the average annual maximum flow is 25,000 cubic meters per second), after which the flow recedes. This is followed by a slight rise in level, then a flow of 2,000–2,500 cubic meters per second, corresponding to the black flood, before a low flow in May of 1,500 cubic meters per second.

The long-term flow data also show deficit periods, including the dry 1970s. From 1980 to 2000, the mean annual flow measured at Onitsha was only 4,720 cubic meters per second, with a volume of 149 cubic kilometers per year; and the lowest flow ever recorded at Onitsha was 109 cubic kilometers in 1984.

The Niger Delta

Rainfall in the Niger Delta is typically 2,700–3,000 millimeters per year over an area of 30,000 square kilometers and actual evaporation is about 1,000 millimeters per year; this means a total runoff of 1,700–2,000 millimeters—and an additional flow of 50–60 cubic kilometers—that is calculated for the water balance of the Niger system. A total of 250 cubic kilometers per year discharges into the Gulf of Guinea.

The Niger River: A Country Perspective

River flows in the various tributaries from the nine Niger Basin countries, based on observations made before 1960 and 1980–2004, are shown in table 3.5. In both periods, the contribution of the Benue is greater than that of the Niger at their confluence at Lokoja. Whereas the hydrographic and hydrologic characteristics of the major reaches of the Niger River are described above, characteristics specific to each country are summarized below:

- Bordering the Niger River for about 140 kilometers, Benin provides about 3 cubic kilometers in an average year from the Mekrou, Alibori, and Sota tributaries originating in the Atakora Massif and Bourgou.
- Burkina Faso's contribution to the river's flow is only 1 cubic kilometer in an average year.
- Cameroon's flow contribution to the Donga, and especially the Katsina Ala, reaches 14 cubic kilometers in an average year. A total of 34 cubic kilometers, from Chad and Cameroon, arrive in Nigeria from the Benue, more than from the Middle Niger River itself.
- Chad's contribution to the Benue is 1.6 cubic kilometers in a normal year; the Benue is the major tributary of the Niger River originating in Central Africa.
- Côte d'Ivoire is estimated to add approximately 4 cubic kilometers in an average year to the flow toward the Niger River, based on runoff of 270 millimeters.

Table 3.5 Average Annual Flow and Flow Volumes in the Niger and Benue Basins from the Headwaters to the Niger Delta, before 1960 and from 1980 to 2004

Country tributary and stations	Surface area (km^2)	Flow before 1960 (m^3/s)	Flow 1980–2004 (m^3/s)	Annual volume before 1960 (km^3)	Annual volume 1980–2004 (km^3)
Niger River					
Guinea					
Tinkisso	—	—	—	220	160
Niandan	—	—	—	260	189
Milo	—	—	—	275	160
Siguiri Station	67,400	1,015	755	—	—
Sankarani				405	265
Mali					
Koulikoro Station	120,000	1,545	1,040	—	—
Côte d'Ivoire					
Bani	—	—	—	670	207
Burkina Faso					
Delta Inflow	222,000	2,195	1,247	—	—
Diré Station	330,000	1,110	750	—	—
Niger					
Niamey Station	—	1,020	670	—	—
Benin					
Malanville Station	440,000	1,140	800		
Nigeria					
Yidere–Bode	—	—	820	—	—
Sokoto	—	—	—	200	100
Jebba Station	1,370	1,600	950	—	—
Kaduna			212	600	400
Baro	730,000	2,525	1,370	—	—
Benue River					
Cameroon					
Riao Station	27,600	280	212	—	—
Chad					
Mayo Kebi	—	—	—	100	80
Cameroon					
Garoua Station	64,000	375	308	—	—
Nigeria					
Gongola	—	—	—	200	120
Taraba	—	—	—	500	380
Donga	—	—	—	500	400
Cameroon					
Katsina	—	—	—	800	675
Nigeria					
Makurdi Station	305,000	3,150	2,380	—	—
Lokoja Station	—	3,400	2,500	—	—
Niger River					
Lokoja Station	—	3,000	1,600	—	—
Onitsha Station	1,100,000	7,000	4,570	—	—

Source: Rodier 1964.

Note: — = not available.

- Guinea remains the "water tower" of the Niger River, with a flow of 36 cubic kilometers in an average year supplied by the Upper Niger at Siguiri, together with the Sankarani tributary.
- Mali, which has a complex water balance, in an average year receives 36 cubic kilometers from Guinea and 4 cubic kilometers from Côte d'Ivoire, then adds 5 cubic kilometers to the river from the Sankarani and various tributaries upstream of Koulikoro and another 10 cubic kilometers from the Bani and minor tributaries of the Dogon region. However, Mali loses 28 to 30 cubic kilometers, 25 cubic kilometers of which is in the Inland Delta through evaporation, before the river reaches Niger.
- Niger's contribution to the river flow is negligible or, more precisely, negative (evaporation). The annual flow increases at Gaya are a result of inflows from Benin.
- In Nigeria, of the 182 cubic kilometers that flow through the city of Onitsha in an average year, only 65 cubic kilometers come from upstream countries, with almost two-thirds of the flow produced within the country itself.

Transport of Suspended and Dissolved Solids

The total suspended solids (TSS) are the particles that are suspended in water. The total dissolved solids (TDS) are the materials that are dissolved in water.

Suspended Solids

This section looks at transport of TSS as it affects the Upper Basin, Inland Delta, Middle and Lower Niger, and navigation on the Niger River. Sediments can be transported by water either in suspension, thereby becoming a part of the water flow because they travel at the same speed as the water, or by being displaced by the current, either by saltation or as bed load at much reduced speeds. Suspended solids tend to be fine sand, silt, and clay, whereas bed load includes heavier sand and larger elements. The amount of solids transported in suspension is more significant than bed load. It is estimated that bed load transport for a system such as that of the Niger River is less than 5 percent of the suspended load.

Upper Basin. The TSS transported in the Niger River result from watershed erosion of streambeds and banks. The annual TSS load carried by the Upper Niger varies between 0.7 million ton and 2 million tons, depending on river flow. Generally, the Upper Niger River has a low sediment load, with average annual TSS concentrations of about 20–30 milligrams per liter, rising to 50 milligrams per liter when crossing dry areas downstream. The Bani River's suspended sediment load is higher, at approximately

50–75 milligrams per liter. Seasonal variations are significant and, with the first floods of the season transporting more sediment, concentrations can reach up to 150–200 milligrams per liter in May–June (300 milligrams per liter on the Bani), dropping to less than 5 milligrams per liter at low flow (figure 3.9). The annual discharge of TSS from 1991 to 1998 was recorded at monitoring stations in the Upper Niger Basin (table 3.6).

Figure 3.9 Flow and Concentrations of TSS for the Niger at Koulikoro [a] and the Bani at Douna [b]

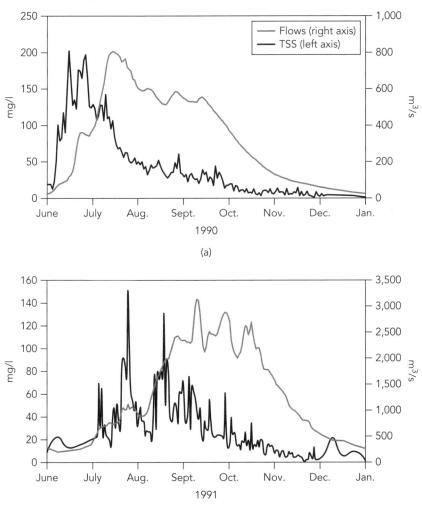

Source: Gourcy 1994.

Table 3.6 Annual Discharge of TSS in the Upper Niger (Banankoro, Koulikoro, and Ké Macina) and the Bani (Douna)

	1991–92	1992–93	1993–94	1994–95	1995–96	1996–97	1997–98
Banankoro							
Flow (10^3 t/yr)	338	411	467	800	862	596	479
Flow (m/s)	531	521	521	1,070	977	825	729
TSS (mg/l)	20.1	25.1	28.4	23.7	27.9	22.9	20.8
Ts (t/km²/yr)	4.8	5.8	6.6	11.3	12.1	8.4	6.7
Koulikoro							
Flow (10^3 t/yr)	607	593	665	1,296	1,014	960	986
Flow (m³/s)	767	775	732	1,480	1,310	1,050	1,019
TSS (mg/l)	25.1	24.3	28.8	27.8	24.5	29.0	30.7
Ts (t/km²/yr)	5.1	4.9	5.5	10.8	8.4	8.0	8.2
Ké Macina							
Flow (10^3 t/yr)	—	715	1,028	1,974	1,701	1,223	1,139
Flow (m³/s)	—	681	647	1,320	1,180	929	911
TSS (mg/l)	—	33.3	50.4	47.4	45.7	41.8	39.6
Ts (t/km²/yr)	—	5.1	7.3	14.0	12.1	8.7	8.1
Douna							
Flow (10^3 t/yr)	257	229	315	729	389	422	448
Flow (m³/s)	190	139	135	459	224	200	202
TSS (mg/l)	42.8	52.3	74.1	50.3	55.1	66.9	70.3
Ts (t/km²/yr)	2.5	2.3	3.1	7.2	3.8	4.2	4.4

Source: Olivry and others 1998.

Note: — = not available; Ts (*transport spécifique*) is the annual discharge of TSS divided by the basin area.

Analysis of the seasonal distribution of TSS shows that, regardless of where the monitoring station is on the river, more than 70 percent of the annual transport occurs in the August–October period of high river flow. Over this period, the significant increase in sediment transport is accompanied by an equally significant decrease in sediment concentration. The monthly average sediment transport figures in August and September are similar. Despite the heavy TSS concentrations reached in the initial erosion phase during July, these concentrations account for only about 12 percent of the annual sediment transport.

Inland Delta. TSS levels decline as the river passes through the Inland Delta, with significant settling in the central lakes and in Lake Debo, notably upstream of Akka. Some increases in TSS occur in the downstream part of the Inland Delta, with the appearance of the first dune belts upstream of Diré. The annual sediment transport flows at the Inland Delta

Table 3.7 Average Annual Flow and Discharge of TSS at Akka and Diré in the Inland Delta

	1992–93	1993–94	1994–95	1995–96	1996–97	1997–98
Akka						
Flow (10^3 t/yr)	591	654	1,033	989	897	832
Flow (m^3/s)	577	571	1,209	892	753	739
TSS (mg/l)	32.5	36.3	27.1	35.2	37.8	35.7
Diré						
Flow (10^3 t/yr)	784	766	1,487	1,183	962	887[a]
Flow (m^3/s)	574	563	1,084	866	745	729[a]
TSS (mg/l)	43.3	43.2	43.5	43.4	41.0	38.6[a]

Source: Picouet 1999.
a. Flow estimated in 1998 by correlation between stations.

Table 3.8 Assessment of the Average TSS in the Inland Delta from 1992 to 1998

thousands of tons

Regions of the Inland Delta	1992–93	1993–94	1994–95	1995–96	1996–97	1997–98
At entry to Inland Delta[a]	944	1,343	2,703	2,090	1,645	1,587
Lake Debo Outlet[b]	696	747	1,250	1,162	1,032	957
Upstream Delta balance	−248	−596	−1,453	−928	−613	−630
Delta Outlet[c]	784	766	1,487	1,183	962	887
Downstream Delta balance	+88	+19	+237	+21	−70	−71
Total balance	−160	−577	−1,216	−907	−683	−700

Source: Picouet 1999.
a. Total flow from the two inlets (Ké Macina and Douana).
b. Total flow of the three outlets leaving Lake Debo (Akka, Awoye, and Korientze).
c. Flow at the single delta outlet (Diré).

stations for the hydrologic years between 1992 and 1998 are presented in table 3.7 of Akka and Diré (Picouet 1999). As with the stations in the Upper Niger, the variations in suspended sediment transport volumes are related directly to river flows. Table 3.8 provides the range of the annual TSS flow balance in different parts of the Inland Delta. The overall balance shows that 0.16 to 1.2 million tons of sediment were deposited within the Inland Delta per year during the observation period (or 17 percent to 45 percent of the TSS entering the Inland Delta).

The Middle and Lower Niger. After crossing the Inland Delta, the Middle Niger River's TSS concentrations tend to increase as a result of by harmattan dust and windblown sand coming from the dunes along the riverbanks. At Tossaye, TSS levels over 100 milligrams per liter can be expected. This trend extends for the entire reach of the Middle Niger. Very erosive floods from the semiarid tributaries in Burkina Faso add suspended solids to these TSS concentrations. For example, the Gorouol, at Dolbel, has an average TSS of 750 milligrams per liter for a flow of 9 cubic meters per second (1976–83, with five months without flow) and monthly concentrations exceeding 1,000 milligrams per liter. The Gorouol transported an average of 180,000 tons in the observation period, whereas the figure for Kandadji was 1.64 million tons.

At Niamey, TSS concentrations again show an increase; data include other years of observation and are not directly comparable to the data for Kandadji, but they show the significance of the Burkina Faso tributaries, the Dargol and Sirba. Over the three years of measurement by Gallaire (1995) at the Niamey station, the annual TSS load was 3.5 million tons, which corresponds to the lowest value of TSS measured at this station. Figure 3.10 shows the lag between peak TSS and monthly flows at Kandadji station.

A significant level of suspended sediment transport by semiarid or dry tropical tributaries also is found in Nigeria, for example, from the Sokoto and Gongola; and on the Benue, from the tributaries of northern Cameroon, where concentrations greater than 10 grams per liter have been measured at the beginning of the season (Nouvelot 1969; Olivry 1978).

TSS loads of the Middle Niger, given the tropical system's tributaries, its violent floods, and its strong erosive power, increase considerably downstream. The Benue has similar characteristics. Although these loads are not comparable to those of some waterways of semiarid areas (such as in North Africa), high TSS loads (the white flood) have begun to pose a silting problem, and bottom dredging must be planned for dams.

According to Meybeck (1984) and Milliman and Syvitski (1992), the Niger transports 40 million tons of sediment per year through Nigeria, as measured at Onitsha. Based on a flow of 154 cubic kilometers in heavy deficit years, the average annual concentration is 260 milligrams per liter. The flow of sediment is 30 percent higher than that of the Congo River, even though the Congo River carries seven times more water to the ocean than the Niger carries.

Impacts on Navigation. The difficulties encountered in river navigation on the Middle and Upper Niger and the upper Benue over the past 30 years give the impression that bed load transport has increased, creating

Figure 3.10 Hydrograph of Average Monthly Changes in TSS at Kandadji

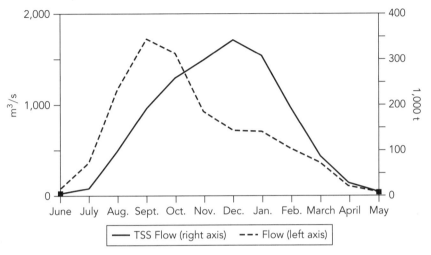

Source: Gallaire 1995.

sandbars and silting up previously navigable stretches. In the Inland Delta and other river areas experiencing progressive narrowing of secondary branches, the drought that dried these channels has not allowed the annual "rinsing" of the seasonal barriers created by deposits of windblown sand and dunes. It will take several years of heavy flooding to reestablish previous water circulation, particularly in the lower part of the Inland Delta. This may lead to the assumption that more sand is being transported, because sandbanks are now seen for six months a year instead of three, and navigation is possible for only four months instead of six. In fact, it is because of low flood levels that the water channel used for navigation is narrower.

Dissolved Solids

This section looks at transport of TDS as it affects the Upper Basin, Inland Delta, and Middle and Lower Niger.

Upper Basin. For the Niger's Upper Basin, TDS transport is 30 percent to 40 percent greater than TSS transport. Between 1.3 and 2.3 million tons of dissolved solids are exported annually to the Inland Delta, depending on river flows; 72 percent to 85 percent of this flow is contributed by the Niger (Ké Macina station) and the remainder by the Bani. This represents a specific flow of 4.7–9.7 tons per square kilometer of catchment area per year. The specific flow is very low at the Douna station on the Bani (from 1.7–6.5 tons per square kilometer per year), compared with that observed at Ké Macina on the Niger (6.2–12.3 tons per square kilometer per year). This latter specific flow is even lower that those calculated at the Niger River stations upstream of Banankoro (8.2–19.6 tons per square kilometer per year) and Koulikoro (8.6–17.5 tons per square kilometer per year).

Inland Delta. The overall transport balance shows that 2 to 5 million tons of solids enter the Inland Delta, half of which are dissolved solids and half of which are suspended solids. The suspended solids portion increases as it travels downstream, where unprotected soils are subject to higher erosion. The TDS difference between stations, and for a single station over different years, is linked to the intensity of drainage and rainfall. The annual balance between entrance and exit of dissolved solids flow over the two largest parts of the Delta (upper and lower) is shown for two contrasting, consecutive hydrologic years, 1993–94 and 1994–95, in table 3.9. A comparison of TSS and TDS for the Upper Basin and the Inland Delta is provided in figure 3.11. Table 3.10 gives an overview of average TDS discharge for 1993–94 and 1994–95.

Middle and Lower Niger. The dissolved solids load shows very little variation in concentrations from upstream to downstream. Table 3.11 gives the average concentrations of major ions over the course of the river. Milliman and Syvitski (1992) cite a value of 59 milligrams per liter, or a TDS load of 9 million tons for a low-water flow year and 10 million tons in

Table 3.9 Average Monthly TSS Concentrations in the Middle Niger

grams per cubic meter

	June	July	Aug.	Sept.	Oct.	Nov.	Dec.	Jan.	Feb.	March	April	May
Gorouol at Dolbel (1976–83)	1,265	868	549	475	604	345	—	—	—	—	—	1,010
Niger River at Kandadji (1976–83)	151	337	229	139	93	46	32	38	46	60	58	66
Niger River at Niamey (1984–86)	184	422	415	356	364	163	108	91.5	87.3	75.8	78	125

Source: Picouet 1999. *Note:* — = not available.

Table 3.10 Average Annual Discharge of TDS in the Inland Delta for Two Consecutive Years

	1993–94		1994–95	
	Volume (km³/year)	TDS flow (10³ t/year)	Volume (km³/year)	TDS flow (10³ t/year)
At entry to Inland Delta	24.7	1,177	56.1	2,367
Lake Debo Outlet	20.8	932	45.7	2,295
Upstream Delta loss	−4.2	−245	−10.4	−73
Diré Outlet	17.8	829	34.2	1,734
Downstream Delta loss	−2.7	−103	−11.5	−560
Total loss	−6.9	−348	−21.9	−663

Source: Picouet 1999.

Figure 3.11 Assessment of the Annual Average Discharge (1992–97) for TDS and TSS from the Upper Niger and Bani Watersheds to the Inland Delta

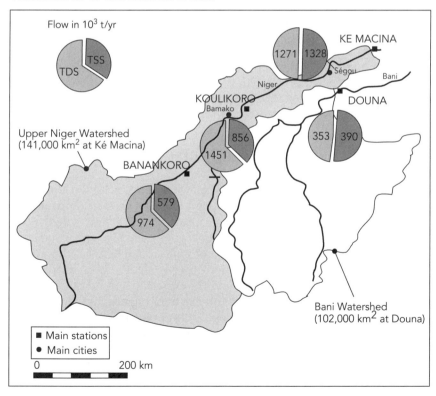

Source: Picouet 1999.

Table 3.11 Average Concentration of Major Ions in the Niger River in Mali and at Onitsha
milligrams per liter

Major ion	Calcium	Magnesium	Sodium	Potassium	Chlorine	Sulfate	Carbon	Total
Niger Ké Macina	2.4	1.1	3.0	1.5	1.0	—	22.3	31.3
Bani	2.8	1.2	2.6	2.1	0.6	—	23.8	33.1
Benue at Garoua[a]	5.6	1.9	3.5	2.0	—	—	30.5	43.5
Onitsha	4.1	2.6	3.5	2.4	1.3	1.0	36.0	50.9

Sources: Picouet 1999 (Mali); Meybeck 1984 (Onitsha).

a. Specific analysis in December 1974.

Note: — = not available.

an average year reaching the ocean (flow of 182 cubic kilometers, which is four times less than the TSS load).

The relative level of total organic carbon (TOC; >0.45 µg, taken in 1996–7 from five upstream Niger River Basin stations, varies from 1.3 percent to 20.45 percent of particle transport. Half of the TOC values were less than 4.95 percent and only 25 percent were greater than 7.1 percent. The higher values are found during flood recession and in low-flow conditions. Concentrations, in milligrams per liter, are most often between 0.1–1.85 (90 percent of the values). On savanna and forested landscapes of the Middle and Lower Niger, the maximum percentage of TOC runoff, which can exceed 30 percent, is seen when TSS concentrations are lower. The annual average is 6 percent to 8 percent of TSS contribution, or approximately 1.5 milligrams per liter.

Water Quality

The degradation of water quality is a significant problem for the Niger. The growth of large cities along the river's banks has not been accompanied by development of wastewater collection and treatment plants, whether for domestic or industrial wastewater. Picouet (1999) establishes a water sampling protocol for the soil and climate conditions found in Africa that allows him to analyze very low concentrations (in parts per billion) of trace elements in the waters of the Upper Niger. The enrichment factor calculated for these trace elements, compared with international standards, allows a preliminary estimate of contamination from anthropogenic sources. The results show that at any of the stations of the upper basin, titanium, aluminum, iron, zirconium, yttrium, strontium, lead, uranium, and vanadium can be considered as earth elements, coming from the weathering of silicate rocks. These elements are correlated to one other and to TSS. Enrichment factors for strontium are very significant at Banankoro, then drop off all the way to Ké Macina. They are the weakest at Douna, which suggests a possible addition of antimony, an element used in metallurgy and other production. Other enrichment factors do not show significant contamination from anthropogenic sources, whether in rainwater or in river water, although water at Koulikoro appears to be more enriched than the water upstream at Banankoro. Koulikoro is located downstream of the city of Bamako, where the main industries responsible for discharging metallic elements can be found (such as tanneries, dye works, galvanizing operations, and textile industries).

Fertilizer use also has an impact on water quality. The impact of cotton farming in southern Mali on wells, pasture, and surface water in the Bani watershed was studied in the 1990s. Traces of pesticides were found, but not in significant amounts. Nitrates, nitrites, phosphates, and ammonia were consistently present at several sites, sometimes over the maximum allowed limits (Bonnefoy 1998).

In the past 40 years, Nigeria has experienced intense exploitation of its petroleum resources, and currently produces 2 million barrels per day from three primary refineries in Port Harcourt, Warri, and the Delta. Petroleum development has contributed to several environmental problems, and in particular those pertaining to water resources. In the Delta, these environmental problems are usually caused by accidental pollution and by illegal siphoning. Studies have indicated that more than 2 million barrels have been spilled into the environment from 1972 to 1982. In addition to petroleum development, the exploitation of coal, iron, gold, and other mineral resources is an environmental threat in the Basin.

Closing Comments on the Technical Chapters

In the preceding chapters, the authors have sought to provide not only a baseline understanding, but also an insight into the unique characteristics and distinctive geographic and geological setting of this "river of rivers" in West Africa. The Niger River is and will continue to be a vital resource to all the people in the Basin. The nine Niger Basin countries, with their specific geographic characteristics and hydrologic contributions, share in the resources of the river. Management of these shared resources calls for a common understanding of and consensus on sustainable use of the Basin's resources. A common understanding of the entire river system will provide a platform for dialogue and cooperative river basin management, leading to shared opportunities and benefits. The final chapter will explore the criteria needed for success in cooperative development of the Niger River Basin.

4
Cooperative Development of the Niger River Basin: Criteria for Success

Promoting Development and Poverty Reduction

It is a basic premise of river basin management that managing the river as a system yields optimal benefits. In the case of the Niger River, this could mean more water, more food, more power, more transport, and so on. Optimized management of any river is difficult, primarily due to the need to recognize so many different interests; with an international river this is particularly difficult, but much can still be done to move in this direction.

The nine Niger River Basin countries are among the poorest countries in the world, and everything possible must be done to improve the lives of the peoples of the Basin. Four of the nine Basin countries are among the bottom 20 countries on the WDI scale, while on the UNDP HDI, seven countries are among the bottom 20. The need for development and investment in the region is evident, and the Niger River holds tremendous potential; this is the reason for cooperation. Development opportunities in the Niger Basin are wide ranging. Some of these opportunities, such as in power, irrigation, and navigation, are directly linked to the river. Once cooperative investments have been made in the development of the water resources, trust and cooperation will grow between the countries, and many other benefits will accrue, include those "beyond the river," such as communication investments, increased trade, enhanced flows of labor and ideas—that is, an enhanced regional integration of the countries of the Basin.

Specific investment opportunities identified by the countries include, but are not limited to:

- Food production. Irrigation potential estimated at 2.5 million hectares, of which only 0.5 million have been developed
- Energy production. Hydropower generation estimated at 30,000 gigawatt hours per year in the Niger River and its tributaries, of which only 6,000 gigawatt hours have been developed

- Access to markets. Transportation facilities, including navigation potential of 3,000 kilometers, of which only 600 kilometers are currently navigable
- Environment. Enhanced environmental management, especially in the Fouta Djallon watershed and the Inland Delta, which will lead to significant benefits for the overall sustainability of the water resources and the Basin
- Flood and drought mitigation. Enhanced flood management, early warning systems, and storage options, which will help reduce the devastating impacts of floods and help further mitigate the impacts of droughts
- Livestock and fisheries. Significant livestock and fisheries potential, currently not commercially developed but highly reliant on predictable water availability
- Ecotourism, which has considerable potential and has yet to be explored

From Unilateral to Cooperative Development

Getting beyond the national agenda is not easy . . . As is to be expected, most countries plan large investments at the national level. When dealing with a shared river basin, and in the absence of an effective basin organization, most countries will plan on a unilateral basis. In some river basins, member states have rushed to "get facts on the ground" through infrastructure projects, seeking to acquire rights ahead of any neighboring states doing the same. In the absence of a cooperative agenda to which countries have committed and which clearly assigns benefits to each member state, the pursuit of unilateral development will most likely lead to lose-lose outcomes, the potential consequences of which include increased tension and insecurity between member states, and lost opportunities for regional cooperation and integration. For many years, the trend in each of the Niger Basin countries has been toward unilateral development of the river's resources. From the position of each nation state, this makes perfect sense, particularly given the lack of a strong regional river basin institution through which cooperative developments could be leveraged, promoted, and instituted.

. . . But it is the only way to secure sustainable win-win benefits. In the Niger River Basin, where water scarcity and flow variability are always causes for concern, the only option for sustainability of the water resources, optimal utilization, and good member relationships is to pursue the path of coordinated, cooperative water resources development. The challenge facing the countries of the Basin is to find ways in which the river's development potential can be realized. With an empowered, enabled, and relevant river basin organization in place, attuned to its constituencies and respected as an institution that can broker major development investments,

the nine countries have an opportunity to move a significant, common development agenda forward, to reduce poverty, promote regional cooperation and integration, and enhance the lives of the 100 million people who live in the Niger River Basin.

Moving from a river basin master plan . . . A common approach to river basin development has been to develop "river basin master plans." This makes sense, because such a comprehensive approach facilitates elaboration of a broad and holistic development plan and provides an overall blueprint for the potential and planned water resources development in the Basin. Donors have at times funded these plans on behalf of river basin organizations in developing countries. Their development is costly, involving many studies and considerable time to complete. Such an approach has merit for a small and relatively simple basin within a nation state. For the Niger River Basin, however—whose watershed encompasses nine countries with a wide-ranging set of investment needs, priorities, and abilities—this approach is of limited real value, because it takes little account of the political and economic reality. Investments will inevitably be driven by a variety of factors, including local and national priorities, diplomacy, political compromise, availability of and access to investment finance, and, perhaps most important, the extent of broad ownership of, and commitment to, defined development priorities.

. . . Toward a more dynamic shared vision of the Niger River Basin. A more dynamic approach to large river basin development will need to be both more pragmatic and attainable. Although detailed analysis and design potential developed through master plans are helpful, they do not contribute directly to building the community of interest and political constituency among and within the member states that is the key to moving cooperative developments forward on Africa's large shared river systems. In several river basins in Africa (and beyond), concerned countries have embarked on the definition of a Shared Vision that encompasses their views of how shared water resources can help in the struggle against poverty, against environmental degradation, for peace, and for regional cooperation and integration. The countries of the Niger Basin have recently endorsed development of a Shared Vision process that lays out cooperative actions through which needs and priorities will be defined and through which management, development, and investment actions will be identified.

Laying the Institutional Foundation for Cooperation

The institutional mandate and its renewal. The 1980 Niger Basin Convention defines the basic scope and mandate of the NBA. This legal framework was established to promote cooperation between the member states and to ensure integrated development in all areas as part of development of its resources, particularly in the areas of energy, water, agriculture, forestry, transportation and communication, and industry. The Convention provides a powerful platform for the NBA to promote, facilitate, and coordinate river basin development among the Basin countries, empowering the institution to play a strong, substantive, and important role in assisting them in their development of the Basin. However, the NBA underwent a crisis during the early 1990s, engaging in a variety of projects and activities that may have allowed for institutional survival but that distanced the institution from its core mandate to manage and develop the Basin. In 2002, the NBA Summit of Heads of State met in Abuja and reviewed the performance of the NBA, agreeing to set the organization on a renewed path toward identifying relevant and strategic priorities, through a Shared Vision process supported by a Sustainable Development Action Program (SDAP), to serve the member countries. The Summit of Heads of State expressed clear expectations of the institution.

At the level of the river basin. During the 1990s, the NBA lost its basic legitimacy, relevance, and constituency, the three key requirements for the viability of the institution. Other important factors, such as capacity and financial viability, are often a result of these three prerequisites.

- *Legitimacy* of the institution is defined by its legal and juridical basis but is also a function of how the institution is perceived by those whom it is intended to serve, its credibility among stakeholders, the level of competency of its staff, and the transparency of its governance.
- The *relevance* of the institution is similarly significant. Does the institution address the real issues faced by the nations in its basin, and thereby serve its constituency? Or is the institution preoccupied with marginal projects or products that do not meet the mainstream development goals set by the countries that are the "shareholders" of the institution?
- It is the "shareholders" of the institution, the *constituency,* who will be the ultimate judge of whether the river basin organization has achieved legitimacy and relevance. Unless the constituency sees the NBA as directly relevant to domestic development priorities, it is unlikely to place real value on the institution. A practical consequence will be that the annual budgets of the institution are left unpaid and it will become increasingly irrelevant to national priorities, initiating a downward spiral that is exacerbated as the institution, to maintain itself, seeks resources from a variety of sources (including donors), involving activities that no longer respect the institution's mandate.

The degree to which the NBA can recapture both legitimacy and relevance will largely determine whether the institution will meet the expectations of its constituency.

At the national level, river basin development is everyone's business. A clear factor for success at the national level is to ensure that the NBA's agenda is owned by many stakeholders. This includes the ministries of water resources and hydraulics as well as the ministries of finance, foreign affairs, energy, agriculture, transport, and environment. In addition, because river basin development is very much a local issue, local governments, basin management agencies, farmers, and communities are equally important stakeholders.

Identifying and empowering a champion. At the national level, therefore, it is important to have a strong champion and coordination mechanism for river basin management. Although "focal points" based in water or hydraulic ministries have an important role to play, they frequently do not have the ability to convene other ministries in key discussions on engagement, priorities, trade-offs, and commitments. A broad national constituency must have ownership of the agenda for national water resources management and development aspirations to be fulfilled on shared water resources. A strong national champion of river basin management—for example, the minister of water and hydraulics or an appointed senior official in the prime minister's office—will be needed to convene opposing interests, to formulate an agenda for engagement, and to forge alliances, both at home and abroad.

The legal framework. An enabling legal framework that is reviewed and updated as needed can go a long way toward facilitating river basin cooperation. That said, a legal framework alone, however solid, cannot be relied on to achieve the institution's development goals. Although the architecture of the legal framework is important—because it provides the mandate and structure of the institution, and as such creates the vehicle with which the cooperation agenda can be driven—it is the driver of the vehicle who will define the path to legitimacy, relevance, and constituency.

Subsidiary agreements may be needed at a later time. In view of the hydrological complexity of the Niger Basin, it is possible that subsidiary agreements among a subset of Basin countries will be required for specific river basin developments, which might not involve, or affect, all the Basin countries. For example, developments on the Niger River tributaries in Chad or Cameroon will have no impact on Guinea, Mali, or Niger and a subsidiary agreement between or among the involved and affected parties would therefore make sense. Such subsidiary agreements should of course fit within the framework of legal principles and agreements that exist at the level of the full river basin and to which the nine Basin countries have agreed.

A Political Mandate: The Shared Vision and Sustainable Development Action Program

Creating an enabling environment for cooperation . . . In Abuja in February 2002, the Niger Basin heads of state agreed to develop a management framework for the Basin through preparation of a Shared Vision and SDAP. The Shared Vision is an expression of the countries' commitment to promote a framework for enhancing cooperation and sharing benefits deriving from the Niger Basin's resources. The Shared Vision process encompasses several objectives. The first objective is *political*, to formulate a statement on sustainable development of the Niger Basin to be adopted by the Niger Basin heads of state. Such a statement must include commitments to, and goals for, cooperation that will lead to joint developments in the Basin. The second objective is *operational*, to prepare the SDAP for the Niger Basin. The SDAP is seen as an appropriate instrument to realize countries' commitment to address the challenges of the Basin. It will include an innovative planning and priority-setting approach to define the development opportunities in which the member countries can jointly participate. The Shared Vision's third objective is *financial*, to mobilize resources from both member countries and international donor partners to implement the SDAP.

. . . Overseen by the Heads of State Summit. To facilitate this process and report on progress to Basin country decision makers, a supervision mechanism has been established. The Niger Basin Council of Ministers reports directly to the Heads of State Summit. The Council is complemented at the national level by National Steering Committees, whose role is to ensure country commitment to, ownership of, and engagement in the proposed decisions and actions that will form the basis for the planning and development agenda of the Niger River Basin.

Making Cooperation Happen in the Niger River Basin

Shared Vision: a phased approach... The Shared Vision process has two main phases. The first phase, nearing completion, includes preparation of national multisectoral studies to explore opportunities and constraints for joint developments in the Basin. Also during this phase, a series of consultations is being held with various stakeholders at the local, national, and regional levels, and with potential donors and sources of technical expertise. In order to develop regional capacity to implement the SDAP, the first phase also includes an institutional audit of the NBA. The goal of this audit is to enable the institution to reform and adapt itself to the challenges of realizing the Shared Vision. The second phase of the Shared

Vision, building on the outcomes of the multisectoral studies, will consist of formulating SDAP actions at the Basin level and identifying specific joint development opportunities within the Basin. This phase of the Shared Vision also includes the creation of an institutional mechanism for the Basin countries to agree on priority actions, based on project location and potential for shared benefits.

Sharing benefits and costs of cooperation. In determining system boundaries for assessment of costs and benefits, it is important to include the entire Basin and to identify all benefits, including those that may be beyond the immediate water resources investment. Environmental benefits (and costs) and direct benefits from water resources investment are potentially the easiest to identify, to cost, and to share. Other less tangible benefits include the reduction of costs that are intrinsic to the absence of cooperation on the river (due, for example, to suboptimal power and food production), as well as those benefits "beyond the river" that will derive from enhanced cooperation, trade, flow of communication, ideas, labor, and so on.[9]

Cooperation options. In West Africa, the experience of the Organisation la mise en valeur du fleuve Sénégal (OMVS), the Basin's management organization, in co-owning and operating assets sets an important precedent for a form of cooperation that is extremely sophisticated. Consequently, other forms of cooperation may be viewed as "less worthy," but this would be a mistake. There are many degrees or types of cooperation on international river basins, with each type defined and shaped by the degree of geographic and hydrological complexity in the river basin, by history, economics, diplomacy, and politics. Figure 4.1 presents a schematic depiction of various types of cooperation in selected river basins.

The Niger Basin states can engage at many levels. As engagement grows stronger, confidence and trust will also grow and a move along the continuum may be both facilitated and appropriate. The Niger Basin countries have committed to cooperation and moving from unilateral action toward enhanced coordination, collaboration, and possibly joint action, even integration. The Shared Vision process and the SDAP will help to achieve these goals.

Joint infrastructure: a primary source of benefits. The obvious form of cooperation is in investing in joint structures for the management of the river, such as in multipurpose storage and river regulation, which can enhance food and power production and lead to greater water security for the countries during drought or water-scarce years. Such investments would be in contrast to the history of unilaterally developed structures, and in contrast to plans for such structures, which meet national needs only and

Figure 4.1 Examples of Types of River Basin Cooperation

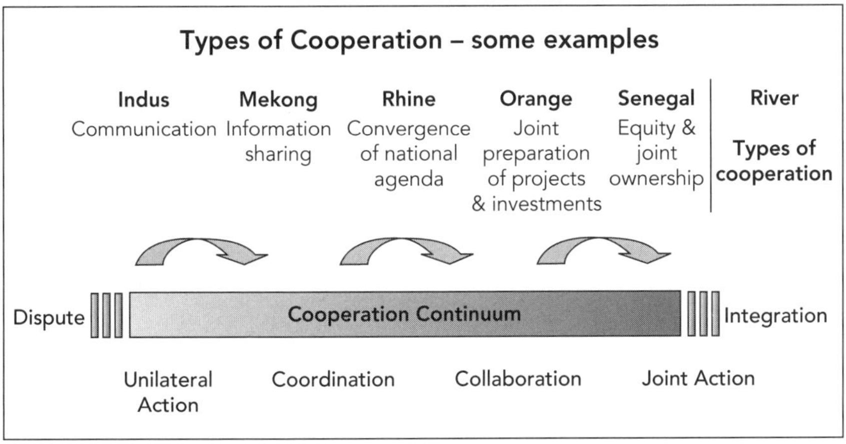

Types of Cooperation – some examples					
Indus	**Mekong**	**Rhine**	**Orange**	**Senegal**	**River**
Communication	Information sharing	Convergence of national agenda	Joint preparation of projects & investments	Equity & joint ownership	**Types of cooperation**

Dispute ||| ⟵ Cooperation Continuum ⟶ ||| Integration

| Unilateral Action | Coordination | Collaboration | Joint Action |

Source: Sadoff and Grey forthcoming.

take no account of the needs of the Basin as a system, or of the Basin population as one community of interest.

Debt sustainability: an incentive for cooperative investment. Some Niger River Basin countries are highly indebted and working hard to manage their debt sustainability. In this context, cooperation on jointly owned and jointly operated works becomes a relevant option. Although the pursuit of debt sustainability alone is an insufficient incentive to make cooperation work, it can, nevertheless, be a contributing catalyst for development of jointly owned assets.

Decision-making tools. To enable the river basin organization to prioritize, plan, and determine optimum investments, the organization must have solid datasets and good river basin models that can facilitate objective analysis of impacts, costs, and benefits. The Niger Basin is in a good position in this regard, because it has one of Africa's best and most sophisticated basinwide hydrological monitoring systems. A next step is to develop tools through which hydrological and economic modeling could be carried out to identify and assess optimum investments.

People and the Environment: A Focus of Cooperation

Engaging stakeholders in issues that affect them. Beyond the obvious areas of economic cooperation in development of the river, there are many ways in which cooperation can improve the lives and livelihoods of the people of the Basin, as well as improve the environment, reversing degradation and enhancing sustainability. The river basin organization has an important role to play in promoting engagement of a wide range of stakeholders with diverse interests. A small and modestly funded organization is limited in what it can do in a basin with about 100 million people—people who speak diverse languages and who are spread across enormous distances. Nevertheless, in a spirit of cooperation and transparency, the NBA can set an example of leadership in a culture of openness, consultation, involvement, and inclusion. This can be done through several media, through interviews in national languages in the domestic media of the member countries and through newsletters and the Web. The NBA, as part of its institutional renewal, intends to follow this path.

Migration: a tradition as old as the river itself. The Niger Basin is at the center of an important migration flow from north to south, due to economic differences and the demand for labor. The poorer and drier countries of Burkina Faso, Chad, Mali, and Niger export their labor to the coastal and wetter countries of Nigeria, Côte d'Ivoire, Benin, and Cameroon, largely for the production of cash crops (including coffee, cocoa, and bananas). To a lesser extent, there is also the migration of the Bozo and Somono fishermen to the large reservoirs on the Niger River and of pastoralists, with their stock, to the Inland Delta in the dry season. Migration and large-scale demographic movements have in large part developed in response to the river and the seasons, with herders moving their cattle and sedentary farmers relying on the bounty of the Inland Delta and flow-recession agriculture to raise their crops. Consequently, migration is long established and accepted in the region and is an efficient means of ensuring the effective, sustainable, and efficient use of natural resources. With population growth that averages 3 percent in the Niger Basin, pressure on existing resources has increased dramatically, in places leading to resource conflicts.

War and conflict add stress. During recent decades, civil war and unrest in several countries that border the Niger Basin have made it the recipient of large numbers of people displaced by violent conflict, adding pressures and stresses to already fragile lands. For example, in 2003 more than 25,000 refugees settled in the Fouta Djallon and Mount Nimba regions of Guinea. This led to increased degradation of the highlands, with rapid deforestation and associated land degradation, soil erosion, gullying, and loss of productive lands. Loss of absorption capacity causes rapid runoff, with the consequences downstream of high sedimentation, siltation of existing infrastructure, floods, and changes in river flows.

Root causes of environmental degradation. Major environmental degradation in the Niger Basin results from either natural or anthropogenic causes.

Natural causes relate to climatic variability and change, in particular the decrease in rainfall in the Basin since the late 1970s. The major anthropogenic causes are land degradation, deforestation, and soil erosion that have taken place in the watershed in large part because of increased demographic pressure.

Four principal environmental issues. Four principal environmental issues—land degradation, water degradation, deforestation, and biodiversity loss—have a synergistic effect on water resources in the Basin. *Land degradation* in the form of erosion results from inappropriate agriculture practices, such as bush fires, clearance for rice paddies, extensive cultivation, overgrazing, and reduction of wetlands from drainage. *Water degradation,* mainly the deterioration of water quality, results from non-point-source impacts from pesticides and fertilizers used in agriculture, from point-source urban pollution, and from lack of sanitation infrastructure (sewerage). *Deforestation* is the result of increased needs for energy and limited access to electricity; people in the Basin use wood and charcoal for domestic purposes (which also contributes to land degradation). *Biodiversity loss* is caused by habitat destruction and a subsequent increase of invasive species, which are in turn caused by inappropriate fishing practices, deforestation, and land conversion for agriculture.

Potential roles of the NBA. Each of these issues is extremely important. However, it is also important that the member countries and donor organizations not burden the NBA with a mandate and expectations of delivering projects and action on the ground in all areas. Nevertheless, whereas resolution of many of these issues clearly lies beyond the capacity of the NBA, the institution can play an important role in increasing awareness of transboundary impacts of socioeconomic pressures on natural resources. In addition, the NBA can assist member countries in identifying investment resources to protect the Basin and its watershed, thereby providing benefits, not only to the country where the protection is undertaken, but also to the downstream countries, which will benefit from the protection of the source through improved water quality, controlled runoff, and reduced sedimentation. In some cases, the downstream benefits from watershed rehabilitation and protection investments upstream might be so significant that downstream countries receiving this benefit will invest jointly in such activities. The subsidiarity principle will help the NBA to identify areas, as part of the SDAP, where the institution will have a comparative advantage over well-established national and local agencies that are also charged with working on these matters.[10] The key question to ask is whether action in this field is part of the core NBA mandate and whether a transnational approach by the river basin organization is the most appropriate way to tackle a particular issue.

Criteria for Success and Ways Forward

There is significant regional development and integration potential among and between the Niger Basin countries, potential that could provide considerable development benefits for all involved. Enhanced cooperation on the management and development of the resources of the river can help to realize this potential, whereas unilateral actions will not. The commitment of the Niger Basin countries to the Shared Vision shows clearly that, at the level of the heads of state, the choice in favor of cooperation has been made. The path ahead is clearly difficult, but, as noted by Nigerian President Olusegun Obasanjo at the Niger Basin Heads of State Summit in January 2004, "The Shared Vision constitutes the ideal, if not the sole, way to follow, so as to preserve the environment and enable joint development of the Niger Basin" (Resolution of the January 2004 Heads of State Summit). It is now up to the NBA and its stakeholders, which include the international donor community, to make this vision a reality. As the countries move forward, the following elements can be summarized as key ingredients for success:

- *Continued strong political leadership* and commitment to the Shared Vision are required to ensure that the process moves forward and that it produces tangible results.
- *Staying the course of the reform process.* The NBA is currently going through a major reform process, which will involve several difficult decisions. However, it is important that the member stakeholders stay the course of reform, so that the institution can regain its relevance and legitimacy, and establish a strong national constituency.
- *Leadership and champions.* Moving a complex, nine-country water management and development agenda forward will take a sophisticated, open, dynamic institution with an inspired leadership, as well as national champions, who clearly see the importance of "hydropolitical" engagement, and who will drive the Niger River Basin development agenda so that the benefits of cooperative investments can be realized, contributing to the poverty-reduction priorities of the member countries.
- *A dynamic and enabled staff . . .* Continued and strong leadership will be provided by the Heads of State Summit. The Council of Ministers, supported by the Technical Committee, will continue to give direction and guidance. However, the day-to-day task of delivering on the Shared Vision for the member stakeholders will be that of the staff of the NBA. It is therefore extremely important that the current reform process take root and that the staff be technically skilled and competent, motivated, and empowered to undertake this vital task. Reform holds the promise of delivering this key ingredient for success.
- *. . . in a financially viable institution . . .* The commitment made at the political level is historic and holds great promise. The instructions given to

the NBA are clear and resounding. Yet, the objectives can be met only if the countries ensure that their financial commitments are kept, allowing the NBA to remain financially sustainable and autonomous, to attract highly skilled staff, and to continue to work on its core mandate of river basin management and development.

- *... that continues to stay on message.* Having a reliable resource flow from the member countries will allow the institution to stay on message and on mandate, because it will not have to pursue marginally relevant projects to ensure institutional survival.

Moving beyond unilateral planning . . . It has been argued in this chapter that cooperative planning and development are the only viable way forward for the Niger Basin countries. The Niger Basin heads of state have made clear commitments to a cooperative path and have instructed the NBA to organize itself accordingly. To manifest commitment to the cooperative agenda, it is important that each of the countries review its national water resources management plans to assess whether viable alternatives might exist if development is now viewed through the lens of the broader, regional river basin.

. . . facilitated by the "hydro-diplomacy" of the renewed NBA. The difficult task of assessing and comparing optimum investments in the Basin, which will yield the greatest number of benefits to the largest number of members, in a context of social and environmental stewardship, is by no means simple. Yet, it is exactly this task that is now required of the renewed NBA if real and tangible development benefits are to be realized for the people of the Niger River Basin.

Development partners commit to their side of the compact. Several donors, including the World Bank, have committed to supporting the member states and the NBA as they embark on reforming the institution and together defining their Shared Vision and SDAP. It is important that the donor community now put aside any individual preferences for national investments, in an effort to let optimal regional solutions emerge through the Shared Vision process. It is equally important that donor partners continue to support the Niger Basin countries and the NBA strongly, as they embark on this historic process that holds the promise of unleashing the development potential of the river for all the people of the nine Basin countries.

Appendixes

Appendix 1: Main Maps of the Niger River Basin

Map 1: Niger River Basin

Map 2: The Upper Basin of the Niger River and the Bani River

Map 3: Inland Delta and Lakes District

Map 4: Middle Niger (Northern Section)

Map 5: Middle Niger (Southern Section)

Map 6: The Benue River

Map 7: The Lower Niger River and the Niger Delta

Map 8: Geographical Features

Map 9: Niger Basin (Lapie, 1829)

Map 10: Navigable Segments of the Niger River

Map 11: Rainfall Patterns

Map 12: Representative Annual Rainfall Distribution

Map 13: Estimated Average Flows and Evaporation Rates on the Niger River

Map 14: HYDRONIGER Data Collection Stations along the Niger River and Its Major Tributaries

Map 15: Interior Delta of the Niger River

APPENDIXES

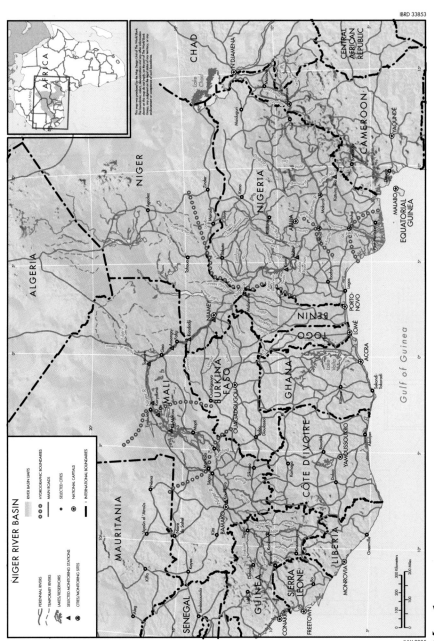

Map 1

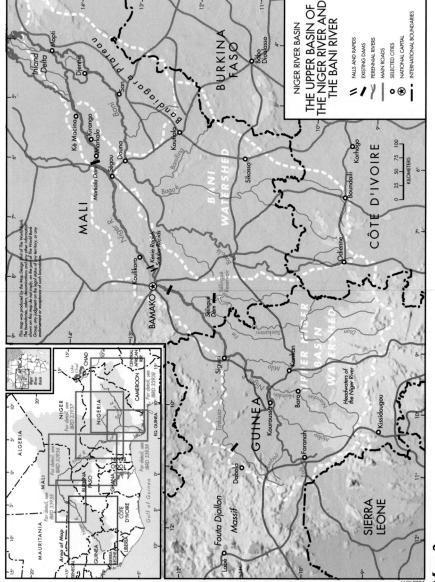

APPENDIXES

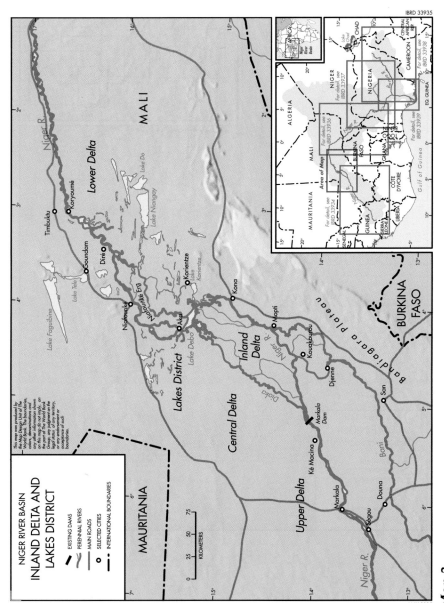

Map 3

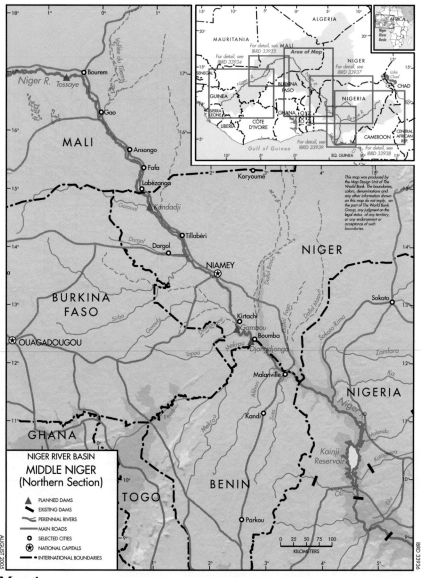

Map 4

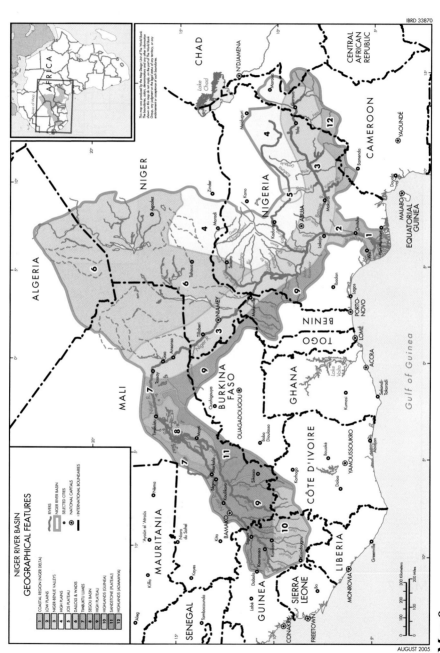

Map 8

APPENDIXES 77

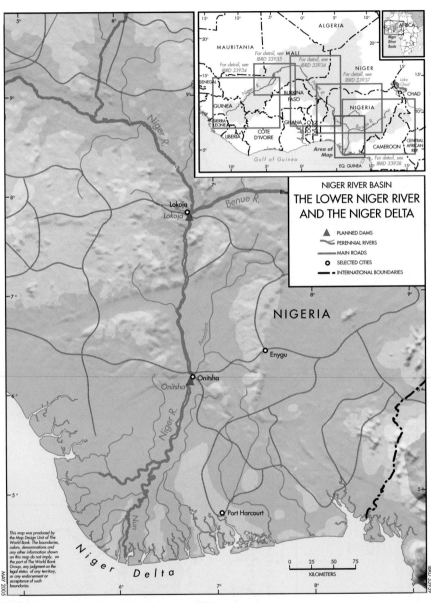

Map 7

76 THE NIGER RIVER BASIN: A VISION FOR SUSTAINABLE MANAGEMENT

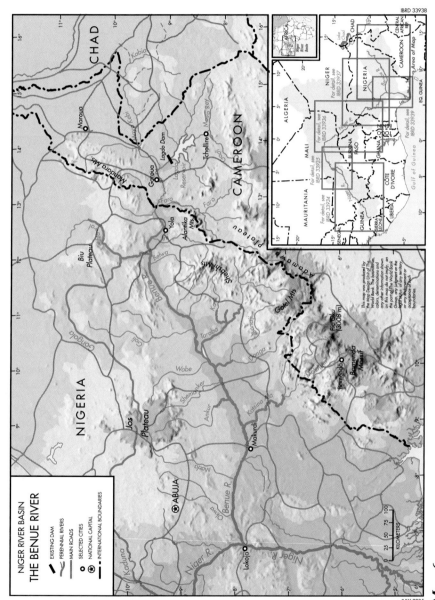

Map 6

APPENDIXES

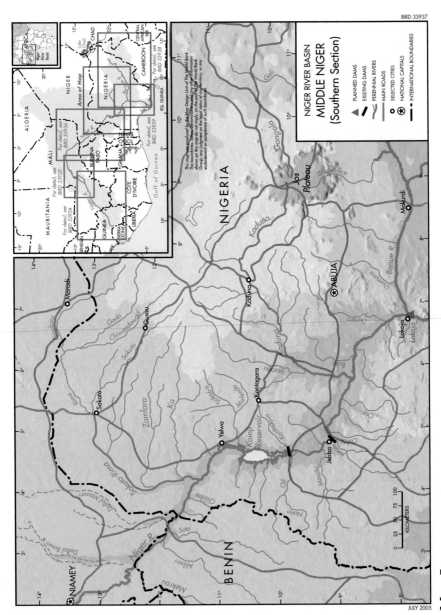

Map 5

APPENDIXES

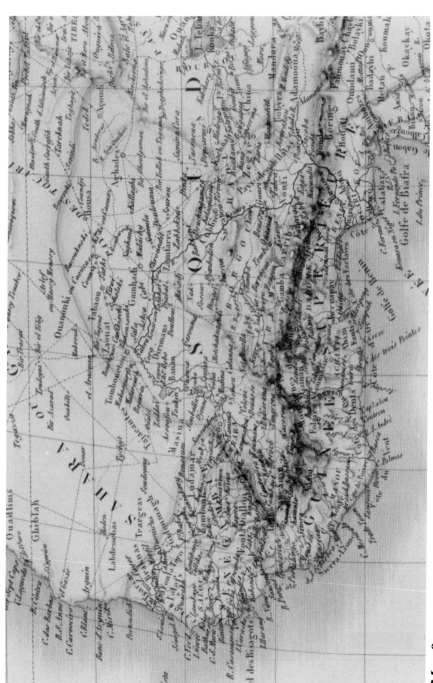

Map 9

80 THE NIGER RIVER BASIN: A VISION FOR SUSTAINABLE MANAGEMENT

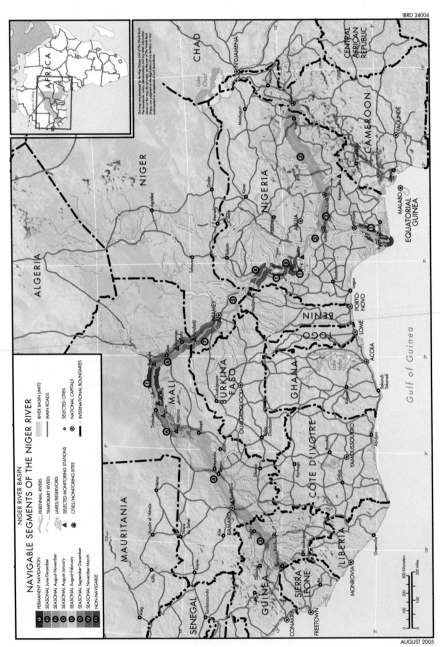

Map 10

APPENDIXES

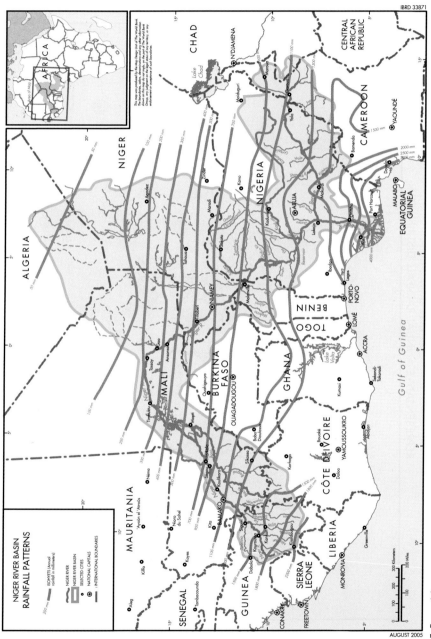

Map 11

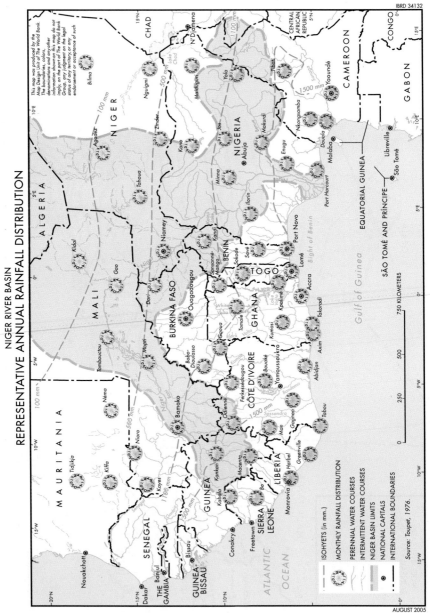

Map 12

APPENDIXES

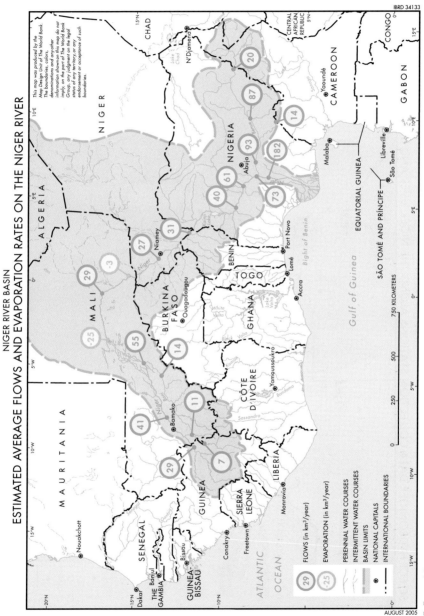

Map 13

84 THE NIGER RIVER BASIN: A VISION FOR SUSTAINABLE MANAGEMENT

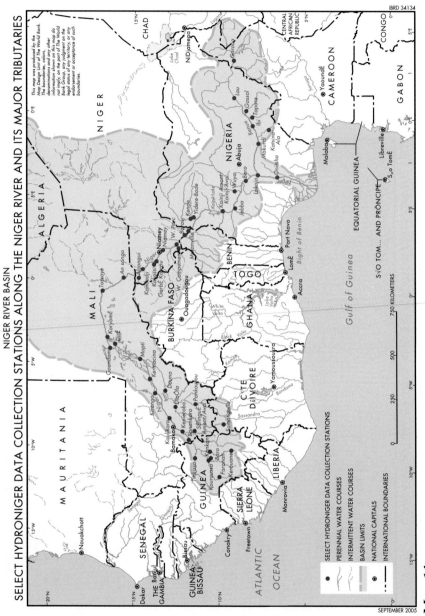

Map 14

APPENDIXES 85

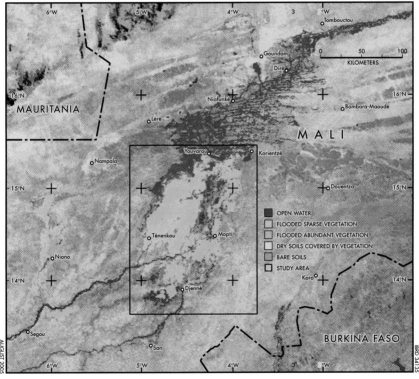

Map 15

Appendix 2: Technical Supporting Information

Table A2.1 Climate Characteristics of the Upper Niger and Middle River

Parameter	Macenta	Guinea Kankan	Siguiri	Bamako	Ségou	Mali Mopti	Timbuktu
T° average year	24.0	26.0	26.9	28.5	28.6	27.7	29.1
T° x months (x)[a]	(3)34.6	(3)36.2	(3)38.0	(4)39.4	(4)41.2	(5)40.0	(5)43.2
T° n months (n)[a]	(12)14.0	(12)14.2	(1)13.8	(1)17.6	(1)15.4	(1)14.0	(1)13.0
Üx year%[b]	96	90	85	73	74	75	54
Ün year%[b]	58	45	39	33	32	31	21
½ (Üx + Ün) % March	69	51	40	26	31	30	21 (April)
½ (Üx + Ün) % August	85	82	81	79	80	78	68
Rainfall avg. (mm)	2,100	1,510	1,250	985	650	415	180
No. of dry months[c]	1–2	4–5	6	7	8	8–9	10

Source: Olivry and others 1995.

Note: T° = temperature.
a. Numbers between parentheses indicate the number of months at maximum temperature (x) and minimum temperature (n).
b. Üx and Ün are maximum and minimum averages of annual relative humidity. (Üx + Ün)/2 corresponds to average relative humidity in the driest month (March) and wettest month (August).
c. According to the definition by Gaussen, a month is considered dry when precipitation < $2 \times T°$ C.

Table A2.2 Annual Evaporation in the Niger Basin
millimeters

Headwaters in Upper Guinea	1,200–1,400
Upper Guinea Plains	1,500
Southern Bani Basin, Sankarani	1,500
Malian Niger Koulikoro-Segou	1,700
Middle Bani Mopti Area	2,000
Niger River Loop, Tossaye, Gao	2,300–2,500
Northeast Burkina, Kandadji	2,350–2,450
Southern Niger, Northern Nigeria, Sokoto	1,900–2,000
Southern Jos Plateau, Adamawa	1,400
Jebba, Baro, Makurdi	1,500
Northern Benue Cameroon, Gongola	1,900–2,000
Onitsha, Lower Niger	1,200
Niger Delta	1,000–1,100

Source: Pouyaud 1986.

Figure A2.1 Histograms of the Average Monthly Rainfall (mm) in the Upper and Middle Niger River Basin

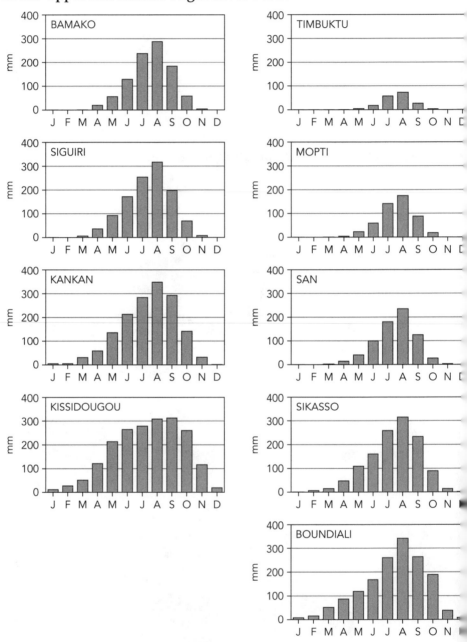

Source: Olivry 2002.

Figure A2.2 Average Monthly Relative Humidity and Evaporation at Kandadji from 1976 to 1983

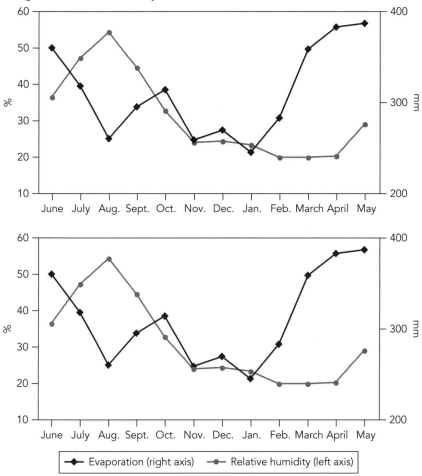

Source: Pouyaud 1986.

Figure A2.3 Seasonal Variations in Monthly Means for Temperature in Mopti and Timbuktu

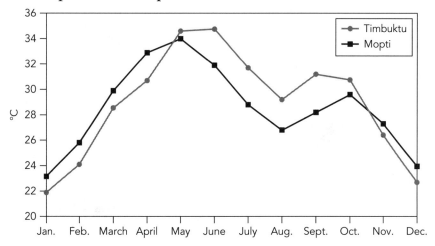

Source: Olivry 2002.

Figure A2.4 Seasonal Variations in Monthly Means of the Relative Humidity in Mopti and Timbuktu

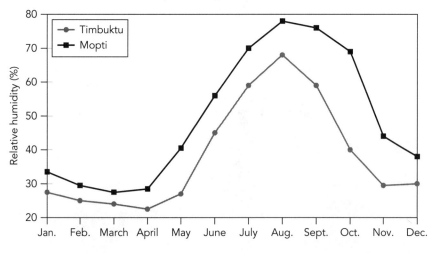

Source: Olivry 2002.

Table A2.3 Hydrologic Parameters Calculated by Decade, 1951 to 1989, at Three Major Stations in the Basin: Koulikoro (Niger), Makurdi (Benue), and Onitsha (Niger)

Balance time frame	Decade 1951–60	Decade 1961–70	Decade 1971–80	Decade 1981–89	Average 1951–80	Average for 1951–89
Niger to Koulikoro						
Flow rate (m^3/s)	1,800	1,600	1,260	795	1,555	1,378
Volume (km^3)	57	50	40	25	49	43
Rain (mm)	1,611	1,529	1,403	1,268	1,514	1,457
Runoff (mm)	473	420	331	209	408	362
Flow deficit (mm)	1,138	1,109	1,072	1,059	1,106	1,095
Flow coefficient (%)	29.4	27.5	23.6	16.5	26.9	24.9
Benue to Makurdi						
Flow rate (m^3/s)	3,294	3,684	3,097	2,609	3,358	3,185
Volume (km^3)	104	116	98	82	106	101
Rain (mm)	1,312	1,294	1,218	1,094	1,275	1,233
Runoff (mm)	347	388	326	274	353	335
Flow deficit (mm)	965	906	892	820	922	898
Flow coefficient (%)	26.4	30.0	26.8	25.0	27.7	27.2
Niger to Onitsha[a]						
Flow rate (m^3/s)	6,771	6,689	5,387	4,629	6,282	5,900
Volume (km^3)	214	211	170	146	198	186
Rain (mm)	1,092	1,011	940	871	1,014	981
Runoff (mm)	194	192	155	133	180	169
Flow deficit (mm)	898	819	785	738	834	812
Flow coefficient (%)	17.8	19.0	16.5	15.3	17.8	17.2

Source: Mahé 1993.

a. The outflow to the Lower Niger can be measured at Onitsha.

Table A2.4 Decrease in Rainfall and Flow over Last Four Decades at Koulikoro and Douna

(a) Koulikoro on the Niger (Watershed area: 120,000 km²)

Periods	Flow rate (Q m³/s)	Rain (mm)	Runoff (mm)	Flow coefficient (%)	Flow index (%)	Rain index (%)	Flow deficit (mm)
1951–60	1,800	1,611	473	29.4	128.6	107.8	1,138
1961–70	1,600	1,529	420	27.5	114.3	102.3	1,109
1971–80	1,260	1,403	331	23.6	90.0	93.9	1,072
1981–89	795	1,268	209	16.5	56.8	84.9	1,059

(b) Douna on the Bani (Watershed area: 101,600 km²)

Periods	Flow rate (Q m³/s)	Rain (mm)	Runoff (mm)	Flow coefficient (%)	Flow index (%)	Rain index (%)	Flow deficit (mm)
1961–70	649	1,187	201	16.9	139.6	106.6	986
1971–80	247	1,053	76.4	7.3	53.1	94.6	977
1981–89	163	945	50.4	5.3	35.1	84.9	895

Source: Mahé and Olivry 1995.

Note: Q = discharge.

Table A2.5 Hydrologic Parameters and Percent Deviation from 1950 to 1969 and from 1970 to 1989 at Select Sites

Niger	Area (km²)	Rainfall (mm)	Flow rate (m³/s) 1950–69	Kf (%)	Rainfall (mm)	Flow rate (m³/s) 1970–89	Kf (%)	Rainfall (mm)	Flow rate (m³/s) % change	Kf (%)
Siguiri	67,600	1,735	1,236	33.3	1,464	755	24.1	−15.6	−38.8	−27.6
Baro	12,770	1,974	271	33.9	1,740	189	26.8	−11.9	−30.3	−20.9
Kankan	9,260	1,974	211	35.1	1,762	160	29.8	−10.7	−24.1	−15.1
Douna	102,000	1,249	685	17.0	1,024	218	6.6	−18.0	−68.1	−61.1
Koulikoro	120,000	1,633	1,719	27.7	1,374	1,048	20.0	−15.9	−39.0	−27.8

Source: Mahé and Olivry 1995.

Note: Kf = flow coefficient.

Table A2.6 Niger River Monthly and Annual Flow at Siguiri from 1950 to 1999
cubic meters per second

Year	Jan.	Feb.	March	April	May	June	July	Aug.	Sept.	Oct.	Nov.	Dec.	Annual
1950	184	111	60.3	43.9	63.2	116	630	1,930	3,770	3,650	1,480	469	1,050
1951	241	145	108	74.9	198	432	1,240	2,680	3,820	3,700	3,640	1,130	1,460
1952	477	266	148	89.4	92.1	185	1,030	2,340	3,400	3,690	1,490	550	1,150
1953	320	176	132	89.9	110	605	1,680	3,300	4,350	3,720	1,540	684	1,400
1954	379	226	146	155	186	528	1,540	2,910	4,530	3,470	2,220	1,130	1,460
1955	483	264	197	148	198	704	1,720	3,120	4,710	4,280	1,970	879	1,560
1956	440	252	166	129	91.6	206	857	1,710	3,640	3,090	1,170	525	1,020
1957	260	134	85.9	48.5	66.3	307	1,190	2,680	4,470	4,680	2,390	794	1,430
1958	400	219	104	117	261	783	967	1,370	3,340	3,190	1,680	960	1,120
1959	403	210	109	65.1	87.5	360	1,270	2,180	4,470	2,760	1,330	519	1,150
1960	247	123	70.8	54.4	85.9	312	1,060	3,050	4,490	3,420	1,530	558	1,250
1961	263	128	66.9	41.8	74.6	107	841	2,270	3,410	2,070	771	308	866
1962	145	72	39.9	32.5	90.1	214	953	2,710	5,300	4,060	1,900	732	1,360
1963	338	208	109	63.5	116	146	669	1,990	3,390	4,110	1,690	522	1,120
1964	239	113	56.8	38.8	48.3	453	1,070	2,430	3,760	3,570	1,140	643	1,130
1965	333	171	108	82.6	89.7	336	1,420	2,040	3,460	2,980	1,220	421	1,060
1966	192	128	84.5	74.5	71.8	164	498	2,200	3,200	3,410	1,570	537	1,020
1967	245	135	91.2	52.4	100	151	766	2,480	4,780	5,740	2,060	696	1,450
1968	350	194	113	81.7	122	769	933	2,520	3,260	2,620	1,270	635	1,070
1969	275	136	87.1	66.1	60.4	306	1,570	2,850	5,040	4,830	3,030	759	1,590
1970	348	176	99.8	76.9	62.7	147	491	1,960	3,540	1,910	828	397	838
1971	160	66.9	42.8	34.5	57.2	133	686	3,120	3,940	2,240	700	440	973
1972	155	78.9	39.4	51	112	214	1,210	1,890	2,830	2,230	925	443	881
1973	163	76.6	35.1	31.3	36	159	326	1,890	2,490	1,480	822	242	648
1974	106	50.6	30.5	22.9	26.5	86.7	1,010	2,440	4,370	3,140	922	308	1,050

1975	128	67.4	30.6	26.5	71.5	183	1,100	2,130	3,910	3,430	1,230	435	1,070
1976	162	78.1	41.3	24.6	65.1	242	681	2,090	2,820	3,650	2,810	741	1,120
1977	313	138	75	35.1	36.3	113	416	1,220	2,400	1,810	664	237	623
1978	108	67.2	45.7	44.7	60.2	385	837	1,730	3,230	2,590	1,090	361	882
1979	162	64.4	29.2	24.8	32.7	317	1,450	3,000	2,940	2,240	1,080	370	982
1980	161	80.2	34.3	16.5	28.9	97	331	1,480	2,540	1,180	816	349	593
1981	135	65.4	26.1	30.2	127	182	948	2,340	2,990	1,990	697	249	819
1982	112	55.2	28.3	29.8	89.6	178	682	1,580	2,260	1,410	721	219	616
1983	94	46.8	26.2	18.5	25.9	295	692	1,490	2,440	1,610	544	194	626
1984	82.5	38.3	21.9	16.4	44.4	125	529	1,710	1,490	1,310	438	160	499
1985	65.6	29.3	11.8	8.08	9.54	22.8	472	1,970	2,940	1,820	491	165	670
1986	61	24.8	8.51	6.11	9.9	32.4	238	1,280	2,760	1,690	614	185	577
1987	72.1	32.7	9.09	4.39	7.85	192	469	1,370	2,090	1,950	673	218	594
1988	78.7	32.6	10.3	6.11	4.52	32.8	346	1,580	2,840	1,130	417	132	551
1989	47.7	19.5	8.65	6.58	7.22	59.6	235	1,090	2,020	1,620	510	204	486
1990	68.8	22.3	9.6	6.93	20.6	64.3	399	1,446	2,144	1,451	538	195	531
1991	73.0	22.1	10.4	8.24	7.25	72.5	437	1,418	2,008	1,587	679	217	545
1992	82.3	32.2	12.1	7.40	6.87	105	567	1,341	2,135	1,379	554	190	534
1993	75.8	25.8	15.4	11.2	13.6	75.5	317	1,494	1,889	1,424	775	281	533
1994	96	34.8	16.8	10.3	10.8	196	731	1,876	3,580	3,218	2,179	531	1,040
1995	190	81.9	32.8	30.0	50.4	134	443	2,209	3,771	3,028	1,179	400	962
1996	155	83.5	31.4	21.4	49.3	156	529	1,752	3,180	2,620	967	313	822
1997	121	55.0	19.9	12.5	30.0	154	711	1,532	2,807	2,068	944	322	731
1998	121	51.4	23.9	12.6	21.0	176	614	2,133	3,062	2,484	881	270	965
1999	106	44.4	19.9	13.3	16.7	49	380	1,370	3,171	2,675	1,445	467	941
Average	200	103	58.6	44.0	67.1	231	804	2,054	3,304	2,708	1,244	454	948

Sources: Mahé and Olivry 1991; Sangaré 2001.

Note: Shaded figures indicate calculated flows.

Table A2.7 Niger River Monthly and Annual Flow at Kouroussa, 1950–2000
cubic meters per second

Year	Jan.	Feb.	March	April	May	June	July	Aug.	Sept.	Oct.	Nov.	Dec.	Annual
1950	34.0	23.0	15.5	13.0	11.0	37.6	162	157	535	649	236	76.6	189
1951	52.5	32.2	24.4	17.4	33.5	94.8	324	527	760	709	916	203	361
1952	86.7	46.8	24.5	13.4	15.8	50.1	230	346	639	751	256	101	243
1953	65.7	32.0	21.7	11.8	18.8	126.2	386	738	988	912	355	151.1	371
1954	106	82.9	70.6	72.0	34.3	109	224	350	703	672	540	292	307
1955	131	69.8	37.1	26.2	35.8	176	347	484	1,121	982	481	266	396
1956	130	85.6	45.6	27.2	20.9	74.2	208	304	561	433	212	105	199
1957	46.3	19.4	10.1	4.6	6.1	44.9	217	486	1,040	1,065	499	194	357
1958	110	56.7	26.7	21.2	39.9	121	138	192	636	774	399	256	260
1959	122	47.3	23.4	7.9	10.7	97.7	225	316	681	429	276	113	218
1960	56.1	20.1	8.3	4.4	7.0	153	349	953	1,098	673	283	119	365
1961	52.4	23.4	11.1	5.0	4.2	5.6	207	531	789	383	194	88.3	222
1962	40.3	17.0	13.5	23.2	50.4	86.6	304	550	928	671	368	179	317
1963	94.0	51.9	22.5	9.6	14.2	16.5	182	446	716	940	353	143	284
1964	69.8	26.8	10.7	5.2	8.5	98.7	236	421	820	808	298	148	286
1965	79.3	33.2	15.0	6.5	8.8	56.3	335	699	1,189	1,187	439	129	406
1966	36.1	24.0	15.5	11.0	8.4	54.3	150	322	613	813	381	225	259
1967	151	44.1	12.1	7.0	10.7	83.5	163	441	953	1,107	369	149	329
1968	76.5	47.3	19.8	10.3	20.1	142	192	494	809	590	256	143	268
1969	65.7	27.8	15.2	9.6	7.8	54.1	251	530	941	1,091	538	180	362
1970	86.3	41.3	18.8	9.73	8.25	17.7	113	235	697	401	214	102	182
1971	39.7	15.7	7.49	4.36	9.87	33.3	151	723	758	480	173	118	246
1972	42.4	16.7	7.51	7.11	12.8	129	341	455	673	530	401	142	270
1973	51.9	14.6	7.45	3.37	5.76	55.8	134	471	542	378	219	83.5	190
1974	30.5	13.2	7.46	4.14	40.4	29.1	226	549	946	767	334	98.0	300

1975	39.3	17.2	10.75	6.33	10.4	32.1	172	566	905	886	652	117	336
1976	53.6	19.0	9.10	4.32	11.3	87.3	171	491	626	820	636	181	304
1977	84.3	30.2	14.00	6.15	8.27	24.3	72.7	255	541	400	143	47	151
1978	31.0	16.7	9.77	6.41	24.0	65.9	173	371	774	612	322	98	246
1979	45.1	16.1	7.26	4.34	5.9	62.6	301	699	650	522	320	100	267
1980	54.7	21.4	8.03	3.97	3.74	27.0	78.8	293	555	266	190	95	152
1981	38.1	20.0	7.08	4.90	30.9	104	282	528	650	541	425	70	264
1982	32.1	14.1	7.12	4.86	15.4	48.8	171	374	513	309	234	63	174
1983	28.7	12.0	5.99	4.02	5.86	69.2	165	316	555	306	147	55.0	163
1984	22.0	10.7	5.76	3.70	7.40	35.9	161	404	358	283	166	48	147
1985	19.1	8.60	4.89	2.49	1.15	2.29	36.5	463	650	414	178	49	180
1986	18.8	7.74	4.11	2.40	2.0	22.5	95.7	306	614	380	208	54	169
1987	14.3	9.40	4.20	2.22	1.7	51.4	263	443	479	447	51.1	62	180
1988	20.0	10.0	5.17	2.44	1.07	3.15	114	374	587	207	94.8	28.3	142
1989	13.0	7.16	3.91	1.52	1.6	27.4	95.2	262	465	362	183	59	146
1990	13.1	8.4	4.1	2.4	2.4	3.3	65.1	259	488	323	133	50.7	113
1991	18.2	7.3	3.5	1.8	1.8	7.2	54.2	198	317	316	130	42.6	92
1992	13.3	6.1	2.8	1.6	3.2	14.5	55.9	247	411	322	125	44.9	104
1993	16.6	11.1	10	5.7	2.2	61.2	37.8	332	387	258	123	41.6	108
1994	13.1	11.9	11.2	4.3	2.2	51.8	149	413	750	838	478	117	237
1995	49.6	24	12.2	7.8	11.2	30.2	104	179	275	253	122	38	92
1996	56.5	29.8	16	9.9	10.9	30.5	136	442	658	576	186	79.5	186
1997	36.9	19.7	10.1	5.8	5.4	31.2	107	217	562	540	259	85.2	157
1998	37	17	8	3.6	3.8	34.7	142	424	547	548	171	59.7	167
1999	25	10.9	5.4	3.6	3.2	12.4	54.9	168	460	533	234	75.9	133
2000	34.4	16.9	8.4	5.2	5.6	37.2	152	387	850	905	350	124	240
Average	52.6	25.4	13.5	8.76	12.3	57.4	180	414	682	595	299	112	232

Sources: Mahé and Olivry 1991; Sangaré 2001.

Note: Shaded figures indicate calculated flows.

Table A2.8 Monthly and Annual Flows of the Niandan at Baro from 1948 to 2000

cubic meters per second

Year	Jan.	Feb.	March	April	May	June	July	Aug.	Sept.	Oct.	Nov.	Dec.	Annual
1948	24.6	13.6	9.8	7.43	47.1	98	372	640	825	459	276	99.2	248
1949	49.1	32	23.7	16.3	15.6	25	93.8	391	907	383	251	93.5	190
1950	33.2	40	25.4	22.7	25.1	39.8	132	318	785	562	244	91.4	193
1951	59.5	26.4	33.5	20.4	39.5	184	488	739	683	853	743	188	338
1952	101	39.1	36.4	22.9	30	76.4	272	582	814	811	352	133	272
1953	37.6	20.3	23.4	22.7	34.5	291	448	743	928	704	365	180	316
1954	95.4	58.1	37.2	59.2	72.8	185	378	512	851	672	533	270	310
1955	126	65.7	71	49	88.3	242	523	731	1,080	732	378	196	357
1956	103	59.7	44.5	41.5	67.9	84.9	326	303	751	511	217	128	220
1957	61.2	33.5	27.8	13.3	30.3	153	499	609	990	857	423	172	322
1958	95.9	52.7	34.6	56.8	127	359	295	215	801	651	472	253	284
1959	109	55.4	29.8	21.3	48.9	158	411	519	866	510	351	136	268
1960	68.9	36.5	19.3	14.8	31.3	147	356	754	902	598	332	129	282
1961	63	35.7	14.5	9.35	31.7	44.9	267	489	481	373	176	67.9	171
1962	31.8	10.2	4.13	4.42	42.2	96.3	198	793	1,150	620	383	193	294
1963	93.1	61.3	31	29	46	48.8	238	404	632	753	253	91.7	223
1964	43.9	20.2	6.23	4.3	20.7	144	309	535	793	565	269	166	240
1965	79.5	53.4	26.7	19.8	37.1	231	500	292	668	447	212	78.4	220
1966	37.0	20.6	14.5	6.76	21.2	115	267	485	429	537	295	91.4	193
1967	43.4	21.9	14.6	6.54	24.8	49.9	182	371	819	894	417	161	250
1968	75	52.2	28.5	23.3	41.4	339	290	603	952	633	351	185	298
1969	87.4	45	32.4	24.1	27.1	139	614	763	943	991	502	162	361
1970	88.2	49.5	35.1	30.4	23.1	47.3	162	326	606	365	206	101	170
1971	44.5	23.4	10.7	10.5	27	110	211	697	808	454	180	123	225
1972	53.3	27.4	10.3	38.9	61.9	248	394	544	666	588	223	129	249
1973	55.1	27.6	8.92	6.31	14.7	32.7	168	375	551	342	221	58.5	159

1974	26.8	12.7	3.87	12.8	8.03	35.3	349	537	978	621	203	77.3	239
1975	36.2	13.5	4.36	2.25	28.7	48.3	183	437	879	638	321	120	226
1976	57.1	30.0	20.8	15.7	36	90.6	145	172	393	779	612	184	211
1977	88.8	48.1	33.0	25.5	45.1	65.2	143	299	583	368	166	56	160
1978	25.5	12.0	8.7	4.85	15.1	190	310	480	713	585	315	113	231
1979	60.4	31.9	13.9	9.32	15.4	166	671	818	664	596	339	122	292
1980	60	33.3	12.9	3.79	4.72	36.9	127	381	502	295	263	111	153
1981	46.3	21	8.48	17.4	68.5	103	348	713	734	559	182	62.9	239
1982	40.7	22.2	8.94	12.2	51.2	103	271	365	454	347	125	56.0	155
1983	25.6	12.0	8.7	9.09	22.5	155	197	430	523	386	173	57.0	167
1984	26.1	12.3	8.9	7.80	20.7	63.6	110	505	295	302	131	28.1	126
1985	11.8	4.2	3.4	3.66	2.03	7.42	131	457	617	258	105	40.6	137
1986	10	2.75	0.853	0.772	0.866	15.6	74	386	576	373	203	45.8	141
1987	20.6	9.2	6.8	6.83	19.4	80.7	184	390	475	430	144	57.0	152
1988	26.1	12.3	8.9	5.75	17.9	11.2	44	387	516	248	113	13.8	117
1989	4.8	0.2	0.7	1.79	12.4	35.6	71	352	442	441	143	30.6	128
1990	12.7	8.7	5.0	3.5	3.5	35.9	57.7	225	423	280	116	45.2	101
1991	20.3	9.0	6.6	5.50	17.5	89.8	129	374	329	463	173	44.3	138
1992	19.8	8.7	6.5	3.61	14.9	89.8	201	376	425	311	173	29.2	138
1993	12.4	4.5	3.6	3.21	14.4	35.9	93	455	421	378	234	56.4	143
1994	25.8	12.1	8.8	3.03	14.1	62.8	109	267	787	729	497	102	218
1995	48.3	25.0	17.4	7.42	20.2	22.4	123	371	488	494	167	33.3	151
1996	14.4	5.7	4.4	3.74	15.1	76.6	113	321	719	888	640	158	247
1997	75.8	40.7	28.0	12.4	27.1	88.0	293	476	804	402	162	33.1	204
1998	14.3	5.6	4.4	2.07	12.8	89.6	211	570	621	495	150	25.3	183
1999	10.5	3.4	2.9	2.61	13.5	7.7	29	133	629	615	261	51.5	147
2000	23.4	10.8	7.8	4.03	15.5	88.6	532	191	317	672	234	42.9	178
Average	49.1	26.3	17.0	14.6	30.5	108	257	464	679	544	282	103	215

Sources: Mahé and Olivry 1991; Sangaré 2001.
Note: Shaded figures indicate calculated flows.

Table A2.9 Monthly and Annual Flow of the Milo at Kankan from 1947 to 2000
cubic meters per second

Year	Jan.	Feb.	March	April	May	June	July	Aug.	Sept.	Oct.	Nov.	Dec.	Annual
1947	45.7	23.1	19.8	6.87	18.6	108	298	453	766	388	125	60.3	193
1948	33.1	14.8	7.69	9.02	33.2	117	419	569	628	267	186	73	196
1949	41	24.7	28.5	34	33.9	37	229	599	763	348	162	70.5	198
1950	52	35.8	21.4	13.1	30	45.2	200	271	533	547	198	73.5	168
1951	45	31.6	28.9	29.5	76.1	142	397	604	641	697	531	146	281
1952	77.3	45.5	36.4	30.6	36.9	73.8	243	598	685	585	221	94.1	227
1953	64.6	31	29.7	19.9	27.4	172	347	595	688	469	198	105	229
1954	61.3	42.1	35.2	49.5	60.9	155	407	599	678	622	379	151	270
1955	80.6	44.3	48.6	46.3	69.9	205	413	598	737	620	237	125	269
1956	68	46.2	44	46.9	39.2	64.4	167	308	509	367	139	77	156
1957	39.6	21.6	20.2	15.6	29.4	87.7	276	471	712	629	244	104	221
1958	55.2	34.2	18	39.4	99.3	275	286	195	605	461	219	132	202
1959	57	30.6	18.2	13.6	32.7	72.6	351	325	787	327	188	73.6	190
1960	34.3	16.9	9.51	19.9	51.4	124	270	720	828	511	219	89.8	241
1961	45.4	25.4	12.2	11.3	34.1	35.1	183	363	484	297	123	53.4	139
1962	28.4	14.8	11.9	20	42.7	72.8	254	459	774	560	268	111	218
1963	53	45.3	32.5	31	55.4	62.2	192	404	662	516	180	68.1	192
1964	32.7	17.8	10.1	10.9	16.8	103	173	498	506	429	164	121	174
1965	65.8	40.5	30	18.4	36.6	136	421	267	565	309	137	56.4	174
1966	16.5	11.8	8.84	11.40	25.8	86.3	232	461	375	391	184	63.7	156
1967	30.5	17.3	13.3	15.5	32.8	55.8	255	534	751	706	227	88.4	227
1968	43.3	26	16.3	21.3	38.2	209	200	530	571	382	210	105	196
1969	59.3	37.7	35.8	31.6	27.8	103	441	757	786	654	355	121	284
1970	67.2	38.7	36.7	41.4	39.7	73.6	167	399	613	280	139	76.4	164
1971	44.3	33.4	21.3	25.9	38.7	54.4	167	455	598	341	115	76.2	164
1972	34.4	20.8	12.2	27.8	59.9	178	292	470	585	359	138	79.7	188
1973	36.4	18.4	9.33	10.8	13.8	44.3	93.2	450	545	303	173	57.1	146
1974	32.1	15.2	10.9	10.1	9.52	27.7	290	467	795	464	145	56.8	194

Year												
1975	29.5	15.8	8.47	21.6	42.5	88.9	238	504	773	498	145	66.8	203
1976	39.1	21.7	12.7	12.5	41.4	66.7	178	374	517	616	412	109	200
1977	62.3	34.4	18	11.5	16.7	51.6	146	274	561	270	97.9	44.7	132
1978	21.2	13.9	9.03	23.8	32.3	152	247	456	654	453	203	73.5	195
1979	43.4	21.1	15.6	25.2	24.9	101	439	693	656	476	205	76.2	231
1980	45.2	32.2	15.9	10.3	24.8	55.1	111	361	470	200	135	64.6	127
1981	34.8	19.8	12.2	15.8	42.5	51.7	357	694	634	279	106	49.7	191
1982	24.4	14.4	8.55	20.7	53	67.6	253	428	518	272	127	47.2	153
1983	23.2	13.2	5.63	5.99	15.2	115	183	406	518	285	103	44.5	143
1984	20.1	8.18	5.49	5.86	22	51.6	123	481	355	215	73.5	29.7	116
1985	10.2	4.06	2.18	2.6	5.09	16.4	170	603	680	332	95.1	35.8	163
1986	14.6	5.98	3.05	3.63	8.75	16.7	97.3	361	556	274	124	38.8	125
1987	17.8	8.99	3.73	2.41	5.39	62.5	174	366	484	322	82.3	44.5	131
1988	15.2	5.15	2.67	1.99	2.18	13.6	76.4	362	513	169	60.8	22.4	104
1989	5.73	3.44	2.18	1.18	4.52	30.8	95.4	327	460	332	81.8	31	115
1990	9.7	3.05	1.25	1.48	15.7	31	101	335	407	220	80.7	33.3	103
1991	14.6	7.4	3.9	2.12	17.2	68.9	136	349	379	350	103	38	122
1992	10.1	4.7	1.49	0.594	6.43	68.9	186	351	448	222	103	30.3	119
1993	9.14	3.62	1.67	7.61	9.1	31	111	431	445	279	146	44.2	127
1994	8.7	2.5	0.8	0.6	1.6	49.9	122	242	707	574	331	67.6	176
1995	19.2	4.5	2.8	7.7	18.1	21.5	132	346	493	376	98.7	32.4	129
1996	10.4	4.9	6	12.3	18.3	59.6	125	296	658	708	432	96.2	202
1997	31.2	8.2	1.7	0.4	9.1	67.6	250	452	719	299	95.5	32.3	164
1998	6.4	2.1	0.3	0.3	13.9	68.7	193	546	588	377	86.9	28.3	159
1999	21.6	9.67	4.93	5.13	8.31	11.2	66.1	107	594	478	165	50.8	127
2000	11.1	2.1	0.1	0.9	7.4	68	417	166	370	526	146	37.3	146
Average	35.2	20.0	14.4	16.0	29.2	81.6	229	439	599	412	179	70	177

Sources: Mahé and Olivry 1991; Sangaré 2001.
Note: Shaded figures indicated calculated flows.

Table A2.10 Interannual Volumes (1992 to 1997) at Main Stations in the Inland Delta of the Niger
cubic meters per second

	May	June	July	Aug.	Sept.	Oct.	Nov.	Dec.	Jan.	Feb.	March	April	Annual
Ké Macina	101	193	548	1,642	3,112	3,076	1,504	597	267	137	92.2	84	948
Douna	2.46	21.3	70.3	455	880	773	382	119	39	17.3	6.66	2.7	231
Nantaka	82	131	357	1,100	1,970	2,130	1,410	756	310	166	114	96.6	701
Akka	101	127	279	802	1,394	1,796	1,968	1,554	884	418	175	107	800
Awoye	0	0	9.31	72.8	158	224	260	192	87.8	27.8	4.34	0	86.3
Korientze	0	0	0	23.4	100	175	176	99.6	44	4.56	0	0	51.9
Diré	78.3	85.7	228	731	1,346	1,686	1,722	1,520	1,001	500	204	99.2	767

Source: Picouet 1999.

Table A2.11 Niger River Monthly and Annual Flows in Mali

cubic meters per second

Year	May	June	July	Aug.	Sept.	Oct.	Nov.	Dec.	Jan.	Feb.	March	April	Annual
						Banankoro							
1991–92	5.3	65.3	415	1,330	1,900	1,590	700	233	87.3	34	9.6	2.5	532
1992–93	4.9	105	543	1,250	2,040	1,360	583	205	79	25.9	14.9	8.5	521
1993–94	12.9	69	297	1,410	1,770	1,410	791	296	105	37.3	17.1	7.1	521
1994–95	9.5	217	704	1,810	3,630	3,390	2,110	549	225	97	42.3	38.1	1,070
1995–96	56.9	141	421	2,160	3,840	3,180	1,170	417	181	99	40	24.6	978
1996–97	55.5	168	506	1,680	3,190	2,740	971	329	136	62.9	22	10.5	823
1997–98	32.5	165	684	1,450	2,780	2,120	949	339	137	58.4	28.2	10.7	730
1997–98	21.7	192	589	2,080	3,060	2,580	890	286	—	—	—	—	—
						Koulikoro							
1991–92	113	215	642	1,470	2,510	2,250	1,020	383	186	137	132	139	767
1992–93	141	264	755	1,430	2,850	2,000	870	383	186	118	125	142	775
1993–94	173	245	522	1,550	2,160	1,980	1,060	47	189	123	131	130	732
1994–95	133	409	1,020	1,940	4,250	5,080	3,050	891	364	219	178	191	1,480
1995–96	197	367	509	2,320	4,920	4,320	1,540	630	323	232	173	170	1,310
1996–97	253	301	526	1,820	3,700	3,660	1,350	448	191	118	98.5	112	1,050
1997–98	163	308	959	1,800	3,800	2,910	1,230	466	195	140	119	143	1,020
1997–98	162	367	772	2,190	4,120	3,910	1,290	477	—	—	—	—	—

(Table continues on the following page.)

Table A2.11 (continued)

Ké Macina

Year	May	June	July	Aug.	Sept.	Oct.	Nov	Dec.	Jan	Feb	March	April	Annual
1991–92	30.7	140	494	1,390	2,240	2,060	1,020	394	160	120	85.5	86.6	686
1992–93	73.3	150	632	1,320	2,540	1,790	818	416	190	78.8	65.9	62.7	681
1993–94	89.5	120	412	1,400	1,930	1,820	1,050	507	197	71.9	75.5	60.6	647
1994–95	61.6	253	954	1,850	3,640	4,530	2,790	851	385	205	138	131	1,320
1995–96	117	264	376	2,040	4,280	4,050	1,530	698	328	218	127	109	1,180
1996–97	164	176	368	1,600	3,170	3,190	1,330	515	235	110	54.5	56.6	914
1997–98	85.3	203	819	1,660	3,300	2,620	1,240	525	214	113	80.5	70	911
1998–99	—	—	—	2,180	3,540	3,400	1,290	—	—	—	—	—	—

Douna

Year	May	June	July	Aug.	Sept.	Oct.	Nov	Dec.	Jan	Feb	March	April	Annual
1991–92	0.1	33.9	50.3	536	839	496	218	64.7	27.5	13.6	4.1	0.9	190
1992–93	0.4	24.9	51.7	228	682	450	152	48.7	20.3	8.6	1.9	0.1	139
1993–94	0	0	88	217	675	420	139	46.6	21.1	8.3	1.6	0.3	135
1994–95	0.1	20.3	115	881	1,400	1,540	1,040	329	85.3	42.1	21.1	10.7	459
1995–96	10.7	33	42.8	442	854	793	345	98.8	40.1	17.3	6.2	1.7	224
1996–97	1.1	28.4	54.1	509	791	663	234	70	28.3	10	2.5	0.7	200
1997–98	6.1	41	74.4	504	948	543	202	65.9	26.5	9.9	3.4	0	202
1998–99	—	—	98	780	1,490	1,430	458	—	—	—	—	—	—

Akka

1991–92	40.2	68.1	224	829	1,330	1,590	1,490	1,030	443	206	115	85.1	621
1992–93	88.5	94.2	277	782	1,210	1,540	1,340	873	395	160	82.2	81.2	577
1993–94	92.2	92.5	216	648	1,230	1,470	1,340	965	453	172	98.1	80.4	571
1994–95	70	128	443	1,060	1,650	2,260	2,930	2,760	1,700	947	379	172	1,208
1995–96	130	174	242	739	1,480	1,940	2,330	1,770	1,070	491	206	130	892
1996–97	124	147	215	780	1,400	1,770	1,900	1,400	802	320	112	70	753
1997–98	75.05	127	380	950	1,410	1,780	1,770	1,280	722	232	81	66	739

Diré

1991–92	30.2	28.9	176	757	1,300	1,560	1,540	1,170	516	246	107	72.3	625
1992–93	66.5	55	213	748	1,180	1,520	1,390	960	449	186	76.3	47.7	574
1993–94	46.7	64.6	153	576	1,190	1,450	1,350	1,020	546	214	86.1	56.9	563
1994–95	42.4	47	366	1,020	1,580	1,960	2,140	2,220	1,850	1,120	468	198	1,084
1995–96	133	138	220	662	1,420	1,800	1,980	1,840	1,230	575	254	134	866
1996–97	103	124	186	648	1,360	1,700	1,750	1,560	932	404	137	59.2	747
1997–98	51.7	93.2	301	847	1,340	1,674	1,660	1,442	808	315	118	93	729

Nantaka

1991–92	50.9	99.1	306	1,150	1,880	1,850	1,080	426	191	139	96.7	82	613
1992–93	77.7	109	366	871	1,680	1,760	817	378	198	104	79	76.7	545
1993–94	81.7	96.3	280	869	1,600	1,620	934	442	206	100	93.2	76.1	536
1994–95	73.7	142	541	1,450	2,450	2,870	2,810	1,780	604	282	175	147	1,110
1995–96	126	209	290	1,160	2,240	2,550	—	754	352	206	128	101	—
1996–97	119	140	237	1,160	2,030	2,400	1,580	580	257	163	—	—	—
1997–98	—	207	576	1,260	2,080	2,310	1,390	573	264	—	—	—	—

(Table continues on the following page.)

Table A2.11 (continued)

Year	May	June	July	Aug.	Sept.	Oct.	Nov.	Dec.	Jan.	Feb.	March	April	Annual
Awoye													
1991–92	0	0	3.2	76	147	188	172	103	28	1.3	0	0	60
1992–93	0	0	9.0	70	129	180	149	82	22	0	0	0	53
1993–94	0	0	2.3	53	132	169	149	94	29	0	0	0	52
1994–95	0	0	27.9	108	198	306	442	406	206	92	20.4	0	151
1995–96	0	0	5.2	64	171	248	320	218	109	33.5	1.3	0	97
1996–97	0	0	2.2	70	158	218	241	158	72	13.8	0	0	78
Korientze													
1991–92	0	0	0	13.6	78	122	90	12	0.5	0.0	0	0	26
1992–93	0	0	0	9.7	57	113	59	9.3	0.0	0.0	0	0	21
1993–94	0	0	0	6.0	24	98	28	15.2	1.3	0.0	0	0	14
1994–95	0	0	0	20.8	144	247	307	266	81	7.2	0	0	89
1995–96	0	0	0	8.5	114	192	236	44	18.6	2.0	0	0	51
1996–97	0	0	0	72.1	164	227	251	164	74.9	0.0	0	0	79
Selingue (dam outlet)													
1991–92	162	189	241	202	542	445	174	124	92.8	108	126	145	213
1992–93	146	175	197	183	720	403	145	134	64	88.9	131	144	211
1993–94	151	168	194	182	383	396	163	121	68.1	91.2	132	151	183
1994–95	148	245	270	203	553	1,116	677	179.2	88.8	118	153	164	326
1995–96	163	211	116	203	945								

Source: ORSTOM/DNHE databank.

Note: — = not available.

Table A2.12 A.B.N/HYDRONIGER—Hydrometry—C.I.P/Niamey, Average Monthly and Annual Flows, Station: 1111500104 Malanville[a]

cubic meters per second

Year	Jan.	Feb.	March	April	May	June	July	Aug.	Sept.	Oct.	Nov.	Dec.	Annual
1970	2,150	2,430	2,140	1,140	365	145	127	643	1,630	1,660	1,460	1,590	1,280
1971	1,740	1,600	796	258	89.5	50.6	216	474	1,390	1,370	1,420	1,630	915
1972	1,750	1,500	652	197	65.4	53.5	117	758	1,320	1,350	1,360	1,460	881
1973	1,330	854	322	98.5	30	18.3	111	563	1,000	1,230	1,340	1,430	694
1974	1,190	561	194	60.8	22.6	19.1	177	621	1,590	1,610	1,510	1,660	769
1975	1,820	1,520	589	155	55.5	31.5	144	663	1,590	1,540	1,540	1,710	943
1976	1,920	1,810	938	271	81.5	69	79.1	474	1,020	1,330	1,470	1,540	915
1977	1,710	1,810	1,300	519	146	116	154	459	1,050	1,060	1,230	1,360	903
1978	1,180	572	201	67	57.9	61.6	92.9	919	1,200	1,280	1,390	1,520	714
1979	1,660	1,460	639	178	53.5	38.9	109	614	1,640	1,660	1,580	1,680	939
1980	1,770	1,440	559	156	45.9	49.4	464	615	1,100	1,170	1,310	1,400	838
1981	1,280	1,020	—	—	—	—	—	—	—	—	—	—	1,160
1982	1,530	898	286	62.4	25.7	—	270	829	1,160	1,140	1,260	1,310	795
1983	1,040	481	232	—	—	—	—	646	889	1,030	1,150	1,180	833
1984	730	281	—	—	—	—	—	428	792	1,120	1,010	911	756

(Table continues on the following page.)

Table A2.12 (continued)

Year	Jan.	Feb.	March	April	May	June	July	Aug.	Sept.	Oct.	Nov.	Dec.	Annual
1985	494	179	50.6	9.29	2.1	5.34	158	749	1,340	1,210	1,250	1,380	572
1986	1,040	394	112	24.8	20.8	14	160	443	974	994	1,150	1,180	543
1987	831	340	109	29	11.3	23	86.4	350	759	1,020	1,050	1,120	479
1988	757	304	87.1	22.4	14.2	46.8	252	1,140	2,100	1,380	1,280	1,340	728
1989	882	337	154	—	—	17.4	148	778	1,180	1,170	1,110	1,150	696
1990	798	305	97.5	25.1	9.36	5.74	75.3	590	1,140	1,050	1,110	1,130	530
1991	707	302	101	31.3	90.8	262	350	968	1,460	1,150	1,180	1,250	657
1992	1,020	445	177	58.6	26.5	43.6	196	782	1,330	1,070	1,130	1,180	622
1993	840	370	127	35	11.1	29.1	132	516	1,070	1,070	1,110	1,150	540
1994	901	447	152	40.6	16.9	95.4	292	1,760	2,260	1,580	1,430	1,490	875
1995	1,620	1,570	1,120	503	264	—	—	787	1,230	1,240	1,270	1,490	1,110
1996	1,550	1,330	782	—	—	—	—	1,020	1,410	1,340	1,220	1,380	1,250
1999	—	—	—	—	—	—	—	—	2,210	1,780	1,440	1,200	1,650
2000	268	162	295	430	565	700	834	1,160	1,130	—	—	—	617
Average	1,440	1,270	979	618	271	147	213	804	1,560	1,500	1,390	1,470	1,030

Source: http://aochycos.ird.ne/HTMLF/ORGINT/HYDRONIG/INDEX.HTM.

Note: — = not available.

a. Niger Malanville. Country: Benin; basin: Niger; river: Niger; area: 1 million square kilometers; altitude: 155 meters; latitude 11.52.00 N; longitude 003.23.00 E.

Table A2.13 A.B.N/HYDRONIGER—Hydrometry—C.I.P/Niamey, Average Monthly and Annual Flows, Station: 1331500034 Yidere Bode[a]

cubic meters per second

Year	Jan.	Feb.	March	April	May	June	July	Aug.	Sept.	Oct.	Nov.	Dec.	Annual
1984	941	426	163	565	464	782	898	456	1,070	1,390	—	1,120	—
1985	664	300	100	384	271	565	390	1,220	1,870	1,590	1,440	1,600	775
1986	1,280	589	246	608	690	763	391	773	1,460	1,470	1,410	1,400	769
1987	1,060	511	210	—	—	—	—	548	1,110	1,340	1,270	1,360	—
1988	1,040	467	164	494	—	—	561	1,800	2,980	2,170	1,600	1,570	—
1989	1,120	507	202	564	388	680	—	1,180	1,750	1,630	1,360	1,360	—
1990	1,000	434	156	439	374	383	182	969	1,510	1,330	1,360	1,360	702
1991	—	—	—	—	—	518	—	—	1,930	1,520	1,440	1,480	—
1992	1,260	627	268	702	—	—	—	—	—	—	—	—	—
1993	845	—	—	—	—	—	—	—	1,540	1,370	1,350	1,380	—
1994	1,110	580	209	439	—	151	458	2,050	2,950	2,540	1,730	—	—
1995	1,870	1,900	1,440	604	285	226	361	1,150	1,610	1,570	1,520	1,680	1,190
1996	1,820	1,610	682	273	169	250	273	1,090	2,100	1,780	1,550	1,600	1,100
1997	1,640	1,100	459	168	—	150	286	912	1,700	1,540	1,490	1,610	—
1998	1,550	966	325	124	117	405	888	2,360	2,940	2,660	1,740	1,770	1,320
1999	1,830	1,510	636	214	122	162	479	1,620	3,170	2,530	1,800	1,780	1,320
2000	1,880	1,840	1,130	355	119	183	421	1,570	1,730	1,710	1,570	1,650	1,180
2001	1,750	1,470	603	181	492	135	718	1,850	2,710	2,110	1,660	1,800	1,250
2002	—	1,400	561	—	—	—	—	—	—	—	—	—	—
Average	1,330	955	444	156	982	178	423	1,300	2,010	1,780	1,520	1,530	1,070

Source: http://aochycos.ird.ne/HTMLF/ORGINT/HYDRONIG/INDEX.HTM.

Note: — = not available.
a. Yidere Bode. Country: Nigeria; basin: Niger; river: Niger; latitude: 11.23.00 N; longitude: 4.08.00 E.

Table A2.14 A.B.N/HYDRONIGER—Hydrometry—C.I.P/Niamey, Average Monthly and Annual Flows, Station: 1331500029 Jebba[a]

cubic meters per second

Year	Jan.	Feb.	March	April	May	June	July	Aug.	Sept.	Oct.	Nov.	Dec.	Annual
1980	1,440	1,640	—	972	—	—	—	—	—	—	—	—	—
1981	—	—	—	—	922	952	1,000	1,000	1,630	1,530	1,000	950	—
1982	—	1,180	—	1,090	—	935	909	1,130	—	—	874	862	—
1986	—	—	—	—	—	—	—	951	687	333	—	829	—
1987	1,100	—	944	747	848	—	839	829	—	—	863	—	—
1988	1,030	706	—	930	565	—	—	—	—	—	—	—	—
1989	—	—	—	—	—	—	364	1,160	1,610	1,850	815	1,110	—
1990	1,050	937	895	221	505	449	149	332	856	455	—	—	—
1991	—	—	—	387	—	—	—	—	—	—	—	—	—
1992	—	—	—	—	—	—	411	209	—	—	—	1,110	—
1993	—	—	—	—	770	612	523	538	1,120	918	530	350	—
1994	435	—	495	529	319	—	—	—	2,310	3,910	1,790	—	—
1995	1,710	2,070	1,760	1,430	1,520	—	—	1,660	—	—	—	—	—
1996	903	823	746	500	557	358	665	702	1,240	1,130	1,070	893	799
1997	1,210	1,230	872	947	513	—	—	—	—	—	—	—	—
Average	1,110	1,230	952	775	724	661	608	851	1,350	1,450	992	872	799

Source: http://aochycos.ird.ne/HTMLF/ORGINT/HYDRONIG/INDEX.HTM.

Note: — = not available.

a. Jebba. Country: Nigeria; basin: Niger; river: Niger; latitude 9.10.00 N; longitude 4.50.00 E.

Table A2.15 A.B.N/HYDRONIGER—Hydrometry—C.I.P./Niamey, Average Monthly and Annual Flows, Station: 1331500002 Onitsha[a]

cubic meters per second

Year	Jan.	Feb.	March	April	May	June	July	Aug.	Sept.	Oct.	Nov.	Dec.	Annual
1980	2,030	2,080	1,290	1,080	1,380	2,590	4,530	10,800	16,500	14,600	6,610	2,600	5,510
1981	1,510	1,200	989	965	1,650	2,490	6,530	10,900	17,300	14,600	4,870	1,890	5,410
1982	1,730	1,590	1,090	1,530	1,690	2,390	5,530	8,060	11,900	12,100	5,570	3,270	4,700
1983	2,270	1,710	1,780	1,800	1,980	2,980	4,230	5,960	10,000	7,490	1,860	928	3,580
1984	714	695	524	699	971	2,320	4,800	7,210	10,600	8,260	3,340	1,450	3,470
1985	923	739	646	856	925	1,770	4,810	9,400	14,900	13,100	4,260	2,040	4,530
1986	1,290	993	897	1,050	1,120	2,000	3,800	7,620	11,400	12,100	5,140	1,910	4,110
1987	1,600	1,480	1,870	2,150	2,450	2,660	3,320	5,420	11,800	12,200	4,630	1,590	4,260
1988	1,090	947	895	1,390	1,450	2,170	3,030	5,990	14,200	16,000	5,230	2,350	4,560
1989	1,620	1,390	1,070	1,460	2,970	3,210	5,710	9,390	15,700	16,200	5,550	2,550	5,570
1990	—	—	—	—	1,990	2,490	5,000	10,400	14,300	12,300	5,390	2,730	—
1991	1,640	1,510	1,200	1,370	1,900	5,840	8,290	13,600	17,600	12,800	6,030	2,640	6,200
1992	1,920	1,600	1,410	—	—	—	—	—	—	—	—	—	—
1995	—	2,580	2,320	2,020	—	—	—	—	—	16,200	—	—	—
1999	—	—	—	—	2,140	2,880	5,920	—	—	—	14,000	4,510	—
2000	—	—	—	—	—	—	—	—	15,500	14,600	5,370	2,950	—
2001	2,780	2,210	1,830	1,760	2,110	2,790	4,660	8,640	13,900	—	—	—	—
2002	—	—	—	—	—	—	—	—	—	—	—	—	—
Average	1,620	1,480	1,270	1,400	1,770	2,760	5,010	8,720	14,000	13,000	5,560	2,390	4,720

Source: http://aochycos.ird.ne/HTMLF/ORGINT/HYDRONIG/INDEX.HTM.

Note: — = not available.

a. Onitsha. Country: Nigeria; basin: Niger; river: Niger; latitude 6.11.00 N; longitude 6.46.00 E.

Table A2.16 A.B.N/HYDRONIGER—Hydrometry—C.I.P/Niamey, Average Monthly and Annual Flows, Station: 1331500007 Makurdi[a]

cubic meters per second

Year	Jan.	Feb.	March	April	May	June	July	Aug.	Sept.	Oct.	Nov.	Dec.	Annual
1980	297	219	189	190	287	1,040	2,490	7,030	10,900	8,530	2,860	726	2,900
1981	305	212	194	204	530	1,010	4,100	6,870	11,300	7,750	2,160	580	2,940
1982	290	223	229	227	350	979	3,460	5,550	8,640	8,080	2,470	668	2,600
1983	346	275	230	226	265	728	2,230	3,940	7,320	3,820	704	294	1,700
1984	219	190	184	211	347	770	2,690	4,810	6,430	4,820	1,520	362	1,880
1985	230	190	174	285	308	1,140	4,080	8,050	9,130	5,720	1,380	474	2,600
1986	271	213	206	240	299	763	2,810	5,270	6,780	5,870	1,920	570	2,100
1987	269	222	212	244	257	766	1,960	3,330	7,150	6,530	1,270	410	1,890
1988	245	202	203	214	291	690	1,800	4,330	9,470	9,640	2,330	601	2,500
1989	272	208	188	203	659	1,370	2,720	6,290	11,100	9,690	2,390	635	2,980
1990	397	263	217	228	474	1,190	3,850	8,140	11,000	7,530	2,790	814	3,070
1991	419	260	232	266	1,120	3,220	4,260	7,760	10,800	7,340	3,010	906	3,300
1992	446	303	262	324	667	1,850	3,840	6,860	10,500	9,760	3,930	1,220	3,330
1993	548	283	—	—	—	—	—	—	—	—	—	1,030	—
1994	511	322	264	265	319	1,220	2,700	5,520	10,700	—	—	—	—
1995	450	289	257	266	602	1,530	3,950	7,980	11,500	11,700	4,490	1,620	3,720
1996	922	617	—	—	—	—	—	—	—	—	—	—	—
1998	—	—	—	—	—	—	—	—	11,300	—	—	2,140	—
1999	1,320	827	775	684	637	1,210	2,870	3,430	—	—	—	—	—
2000	—	—	—	—	—	—	—	—	—	—	—	—	—
2002	—	297	269	—	—	—	—	—	—	—	—	—	—
Average	535	338	287	342	615	1,560	3,440	6,230	10,100	9,550	3,080	998	2,920

Source: http://aochycos.ird.ne/HTMLF/ORGINT/HYDRONIG/INDEX.HTM.

Note: — = not available.

a. Makurdi. Country: Nigeria; basin: Niger; river: Benue; latitude 7.45.00 N; longitude 8.32.00 E.

Table A2.17 A.B.N/HYDRONIGER—Hydrometry—C.I.P./Niamey, Average Monthly and Annual Flows, Station: 1331500014 Lau[a]

cubic meters per second

Year	Jan.	Feb.	March	April	May	June	July	Aug.	Sept.	Oct.	Nov.	Dec.	Annual
1980	124	87.2	60.2	48.3	100	362	1,360	3,230	—	—	—	—	—
1981	147	77.7	43.7	33.3	83.1	263	1,460	2,530	—	2,260	1,120	716	—
1982	371	41.5	25.0	19.1	36.3	380	1,470	2,320	—	—	884	416	—
1987	—	—	—	—	—	—	—	836	1,760	1,170	222	134	—
1989	—	—	—	127	225	—	—	—	—	—	—	—	—
1991	130	104	102	84.4	395	685	—	3,140	—	—	—	—	—
1992	—	—	—	—	—	—	—	—	—	—	—	—	—
1994	—	—	—	—	—	—	—	2,870	—	—	—	—	—
1999	—	—	—	—	—	—	—	2,770	—	—	—	—	—
2000	—	—	—	—	—	1,030	—	3,260	3,260	2,200	991	—	—
2001	544	498	—	—	—	—	—	—	—	—	—	—	—
Average	263	162	57.7	62.4	168	544	1,430	2,620	2,510	1,880	804	422	—

Source: http://aochycos.ird.ne/HTMLF/ORGINT/HYDRONIG/INDEX.HTM.

Note: — = data not available.

a. Lau. Country: Nigeria; basin: Niger; river: Benue; latitude 9.12.00 N; longitude 11.16.00 E.

Table A2.18 A.B.N/HYDRONIGER—Hydrometry—C.I.P./Niamey, Average Monthly and Annual Flows, Station: 1051500020 Garoua[a]

cubic meters per second

Year	Jan.	Feb.	March	April	May	June	July	Aug.	Sept.	Oct.	Nov.	Dec.	Annual
1930	—	—	—	—	—	—	—	—	1,970	—	—	—	—
1931	—	—	—	—	—	—	—	1,760	2,900	—	—	—	—
1932	—	—	—	—	—	—	—	—	2,430	—	—	—	—
1933	—	—	—	—	—	—	—	—	2,350	—	—	—	—
1934	—	—	—	—	—	—	—	2,000	1,290	—	—	—	—
1935	—	—	—	—	—	—	—	1,400	2,720	—	—	—	—
1936	—	—	—	—	—	—	—	—	2,540	—	—	—	—
1938	—	—	—	—	—	—	290	680	—	—	—	—	—
1939	—	—	—	—	—	—	275	949	1,120	—	—	—	—
1941	—	—	—	—	—	—	—	—	1,880	—	—	—	—
1942	—	—	—	—	—	—	—	1,100	—	—	—	—	—
1943	—	—	—	—	—	—	187	—	2,140	—	—	—	—
1944	—	—	—	—	—	—	119	711	1,230	497	—	—	—
1945	—	—	—	—	—	—	190	586	1,810	700	94.1	24.1	—
1946	7.70	2.53	1.25	—	—	79.6	311	666	2,280	2,330	241	89.1	—
1947	—	—	—	—	—	50.8	233	1,330	1,940	532	79.2	6.12	—
1948	—	—	—	—	0.967	142	364	2,460	2,580	910	130	29.4	—
1949	6.61	1.28	0.322	0	6.00	29.9	241	1,010	1,100	481	—	30.9	—
1950	11.8	2.85	0.516	—	25.9	50.5	192	973	1,640	695	179	62.4	—
1951	23.2	7.71	1.61	0.800	28.2	40.2	199	956	1,490	—	—	—	—
1952	—	6.86	1.12	0.733	10.4	55.5	199	715	1,320	1,010	174	70.3	—
1953	29.5	14.1	6.00	1.76	35.8	79.5	401	771	1,270	566	88.3	30.2	274

1954	14.3	5.57	1.61	1.00	8.19	104	—	592	1,930	1,060	215	95.1	—
1955	—	21.0	4.58	1.79	10.1	100	366	1,380	2,350	1,530	353	149	—
1956	47.8	25.2	13.7	5.96	3.12	39.0	252	966	1,880	1,150	161	78.7	385
1957	36.5	20.5	11.4	3.46	14.0	161	399	1,040	1,690	1,070	187	67.9	392
1958	35.2	21.1	9.09	1.96	17.8	85.5	307	703	1,400	681	113	50.9	286
1959	30.8	18.6	7.77	2.26	27.9	124	232	457	2,180	626	168	41.9	326
1960	25.7	11.2	1.83	.466	15.9	76.0	595	1,870	2,820	1,190	278	117	583
1961	37.6	10.6	2.06	.333	.322	59.4	757	1,010	2,530	576	152	40.5	431
1962	14.1	2.78	.483	0	.612	99.1	202	1,140	2,350	1,080	157	73.7	427
1963	28.2	11.6	2.38	1.00	11.2	33.1	367	1,760	1,840	958	222	56.3	441
1964	24.4	12.0	3.54	10.6	13.7	49.5	283	680	1,710	850	187	71.4	325
1965	33.6	—	6.12	1.29	2.70	98.4	306	1,770	1,590	431	82.3	30.2	—
1966	15.1	6.10	2.12	1.76	35.1	158	234	913	2,380	555	185	57.8	379
1967	29.4	15.4	4.61	2.86	5.00	39.3	355	734	1,430	597	88.8	50.9	279
1968	24.4	11.0	3.22	1.00	10.1	119	438	1,080	1,800	514	90.6	37.0	344
1969	16.6	8.07	3.12	4.16	16.5	95.0	414	1,910	2,260	1,210	289	82.4	526
1970	31.6	13.4	3.48	2.29	1.58	23.0	202	1,550	2,320	896	242	78.2	447
1971	26.6	11.3	3.35	1.16	0	15.7	275	1,070	1,660	300	—	—	—
1972	15.0	7.58	2.96	1.53	9.80	128	256	584	516	318	69.4	25.9	161
1973	9.74	2.35	.645	0	3.67	38.4	245	1,250	1,340	345	58.1	14.1	276
1974	6.61	2.17	.741	0	10.0	6.86	222	935	1,040	793	—	—	—
1975	11.9	5.78	2.12	.433	2.25	—	290	1,630	2,350	896	122	—	—
1976	—	16.7	6.58	—	—	45.5	298	929	689	672	234	—	—
1977	—	9.69	3.86	.299	—	—	277	992	1,510	254	27.9	9.21	—
1978	7.45	2.32	—	—	41.0	81.9	382	1,090	2,110	624	185	77.6	—
1979	—	8.28	—	—	—	87.1	317	814	710	256	70.4	15.7	—
1980	12.5	5.27	2.99	1.78	12.6	64.1	475	1,640	1,410	504	136	39.4	359

(Table continues on the following page.)

Table A2.18 (continued)

Year	Jan.	Feb.	March	April	May	June	July	Aug.	Sept.	Oct.	Nov.	Dec.	Annual
1981	16.5	6.48	1.60	0.013	0	0.245	310	657	1,300	366	—	—	—
1982	9.14	—	—	—	—	—	100	416	275	126	21.0	4.52	—
1983	0	0	0	0	20.1	54.9	106	241	371	95.3	69.3	68.8	85.5
1984	66.8	71.9	83.1	80.5	79.9	79.9	111	118	159	111	77.8	67.3	92.2
1986	66.0	59.8	78.8	87.1	100	108	160	291	458	217	87.1	74.4	149
1987	69.8	73.5	77.2	83.3	88.6	119	102	175	206	101	65.4	73.6	103
1988	—	—	—	—	—	—	—	884	1,380	931	—	—	—
1989	79.3	71.4	71.8	75.5	59.7	71.5	101	746	599	155	86.0	68.6	182
1990	62.8	57.1	60.0	66.2	—	66.9	283	964	368	128	82.9	73.9	—
1991	66.5	73.6	80.4	80.7	81.8	99.0	124	1,200	894	175	135	118	261
1992	—	1,160	673	—	—	—	—	—	—	—	—	—	—
1993	84.2	82.8	80.6	74.2	74.2	75.1	108	314	296	125	76.3	68.8	122
1994	62.8	60.6	65.5	66.4	67.0	108	215	810	1,870	586	169	105	349
1995	86.0	80.6	75.2	70.7	76.3	105	470	1,740	732	276	137	97.2	329
1996	84.0	76.1	77.4	82.8	72.5	105	326	506	949	471	151	—	—
1997	97.4	131	134	138	160	160	872	1,180	378	199	—	—	—
1998	—	82.3	83.0	84.7	91.4	97.4	169	930	1,690	857	154	119	—
1999	92.5	95.1	105	98.4	133	—	—	—	—	—	—	—	—
2000	—	—	—	—	—	—	—	—	—	—	—	175	—
Average	36.2	52.2	39.6	26.5	33.8	78.9	287	1,010	1,550	627	142	64.0	308

Source: http://aochycos.ird.ne/HTMLF/ORGINT/HYDRONIG/INDEX.HTM.

Note: — = not available.

a. Garoua. Country: Cameroon; basin: Niger; river: Benue; area: 64,000 square kilometers; altitude 174 meters; latitude 9.17.00 N; longitude 13.24.00 E.

Table A2.19 A.B.N/HYDRONIGER—Hydrometry—C.I.P./Niamey, Average Monthly and Annual Flows, Station: 1051500021 Cossi[a]

cubic meters per second

Year	Jan.	Feb.	March	April	May	June	July	Aug.	Sept.	Oct.	Nov.	Dec.	Annual
1954	—	—	—	—	—	—	—	30.0	54.2	—	93.6	—	—
1955	29.9	13.5	—	—	—	76.1	126	319	495	210	168	113	—
1956	—	—	—	—	0	40.8	107	313	363	152	90.0	81.8	—
1957	39.7	19.7	—	2.66	19.2	64.9	178	307	321	200	82.2	44.2	—
1958	27.8	13.3	6.74	1.06	15.9	78.4	109	323	299	111	43.8	29.3	88.2
1959	13.8	3.78	0.580	0	31.3	90.5	119	142	473	129	168	78.9	104
1960	33.8	12.2	1.90	0	15.3	68.9	265	410	579	139	127	96.5	146
1961	48.4	19.9	6.32	0.966	0.161	65.9	237	204	589	160	101	46.1	123
1962	16.2	2.50	0	0	0	62.9	109	277	393	169	81.0	94.3	100
1963	49.7	21.1	4.32	0.366	11.8	57.7	125	556	334	310	111	51.7	136
1964	24.1	13.4	4.67	16.6	13.2	76.3	106	150	—	94.2	54.5	52.3	—
1965	26.9	12.1	4.58	0.366	0	30.3	179	491	308	101	16.1	1.29	97.6
1966	0	0	0	0.266	7.45	37.1	82.7	268	322	121	121	51.4	84.2
1967	17.2	4.50	1.61	0.300	0	18.2	200	233	322	83.1	37.4	31.6	79.1
1968	15.9	6.41	2.70	1.36	2.87	108	164	243	296	73.5	20.8	13.8	79.0
1969	6.32	3.60	2.12	1.20	0.032	57.2	140	424	504	408	201	89.4	153
1970	42.7	14.8	5.51	1.56	3.90	29.8	79.3	339	425	523	198	77.9	145
1971	27.3	12.4	5.41	1.70	0.354	27.2	151	279	398	103	60.9	27.4	91.1
1972	14.0	7.24	2.83	0.766	18.7	125	150	266	200	128	30.5	16.7	80.0
1973	—	—	—	0.733	0	25.9	193	463	341	88.7	21.8	7.09	—
1974	—	1.07	—	—	3.90	5.33	79.7	204	224	103	—	—	—
1975	3.09	1.21	0	0	11.0	35.2	162	393	537	103	59.2	73.8	115
1976	35.2	12.7	4.67	1.96	14.4	10.5	125	215	147	175	40.2	20.9	66.9

(Table continues on the following page.)

Table A2.19 (continued)

Year	Jan.	Feb.	March	April	May	June	July	Aug.	Sept.	Oct.	Nov.	Dec.	Annual
1977	9.13	6.32	4.13	2.41	2.02	7.65	89.5	368	290	29.8	6.46	4.42	68.3
1978	2.74	2.00	1.41	1.70	8.70	19.4	187	249	208	132	97.3	55.3	80.4
1979	14.9	6.19	2.41	0.591	7.00	41.8	135	233	149	—	—	—	—
1980	0.813	0.372	0.301	—	—	—	155	496	278	74.0	—	—	—
1981	6.85	2.58	1.66	2.20	—	—	121	157	196	43.4	19.4	10.2	—
1982	4.55	1.56	0.620	0.593	—	8.84	88.1	326	—	—	10.5	3.16	—
1983	1.16	1.18	0.398	0.351	0.174	33.2	—	—	—	29.0	—	—	—
1984	1.93	0.630	0.459	0.432	4.89	10.4	—	46.5	85.5	30.3	6.62	2.82	—
1985	0.575	—	—	0.331	0.257	24.0	163	273	166	22.4	6.42	2.76	—
1986	—	—	—	—	—	—	105	255	401	71.5	23.3	10.1	—
1987	5.62	2.07	0.524	0.402	0.435	26.1	19.6	152	114	44.8	31.1	22.1	34.9
1988	—	—	—	—	—	—	139	—	524	134	—	—	—
1989	—	—	—	—	—	12.3	32.8	181	143	37.4	13.4	6.80	—
1990	2.02	0.594	0.292	0.119	0.501	30.3	211	—	118	—	—	—	—
1991	—	—	—	—	—	—	—	—	—	—	—	—	—
1992	30.7	—	—	—	—	—	—	—	—	—	—	—	—
1995	5.00	1.19	0.522	0.258	2.45	—	—	—	—	1320	—	—	—
1996	—	1,400	1,420	1,460	1,520	1,490	1,580	1,590	1,120	—	—	—	—
1999	—	—	—	—	2.57	11.7	81.4	225	347	227	89.6	64.7	—
2000	42.0	23.1	11.6	5.70	5.88	56.9	153	313	166	43.8	12.6	21.8	71.3
Average	**18.2**	**49.8**	**49.9**	**47.1**	**52.3**	**84.7**	**174**	**317**	**331**	**165**	**68.0**	**40.7**	**97.2**

Source: http://aochycos.ird.ne/HTMLF/ORGINT/HYDRONIG/INDEX.HTM.

Note: — = not available.

a. Cossi. Country: Cameroon; basin: Niger; river: Mayo Kebi; area: 26,000 square kilometers; altitude: 192 meters; latitude 9.36.00 N; longitude 13.52.00 E.

Table A2.20 A.B.N/HYDRONIGER—Hydrometry—C.I.P/Niamey, Average Monthly and Annual Flows, Station: 1051500023 Riao[a]

cubic meters per second

Year	Jan.	Feb.	March	April	May	June	July	Aug.	Sept.	Oct.	Nov.	Dec.	Annual
1950	—	—	—	—	5.93	30.7	114	954	1,560	—	—	17.0	—
1951	9.80	3.92	0.935	—	—	28.7	135	701	1,110	618	140	26.5	—
1952	21.9	12.2	6.58	0.166	12.0	27.8	237	712	—	—	—	—	—
1953	8.25	3.39	1.35	0.266	10.9	41.2	220	666	745	333	39.9	9.38	173
1954	—	—	1.32	1.00	2.29	31.1	—	569	1,520	—	97.8	23.6	—
1955	6.80	2.42	—	—	—	42.7	283	1,030	1,810	1,130	161	51.9	—
1956	22.5	—	—	—	—	—	235	673	1,490	921	101	35.9	—
1957	13.6	5.21	1.48	—	5.93	112	265	752	1,240	765	90.1	13.7	—
1958	1.54	—	0	0	0.096	13.3	194	485	936	395	43.1	5.19	—
1959	0.903	0	0	0	1.12	17.7	84.1	327	1,730	264	25.0	—	—
1960	—	0	0	0	2.83	26.1	328	1,220	1,990	827	96.5	14.3	258
1961	4.61	0.964	0	0	0.290	34.1	485	794	1,460	275	29.9	7.41	298
1962	1.77	0	0	0.033	2.16	39.9	82.8	921	1,780	662	70.4	14.6	269
1963	5.29	0.964	0	0.466	4.83	9.86	269	1,180	1,090	554	88.1	21.4	244
1964	10.3	5.55	2.67	5.33	11.0	26.2	205	538	1,440	541	108	30.0	212
1965	12.1	7.67	4.16	2.26	4.80	37.9	143	962	1,070	249	40.4	12.7	268
1966	5.77	2.60	0.870	0	14.4	86.6	134	805	1,710	367	72.4	15.6	176
1967	7.48	3.96	1.93	1.50	3.90	19.3	187	475	947	412	40.2	11.6	231
1968	5.54	3.00	1.22	0.400	3.61	11.9	286	897	1,150	351	45.2	16.4	347
1969	8.67	4.82	2.00	4.53	11.8	56.9	294	1,410	1,590	668	88.8	18.6	—
1970	7.77	4.10	1.54	0.633	.419	10.8	159	1,360	1,660	—	—	—	206
1971	9.90	5.17	2.64	0.766	0	13.0	198	844	1,190	168	28.6	9.16	

(Table continues on the following page.)

Table A2.20 (continued)

Year	Jan.	Feb.	March	April	May	June	July	Aug.	Sept.	Oct.	Nov.	Dec.	Annual
1972	5.25	3.24	1.19	1.29	5.29	29.0	133	385	328	239	27.9	6.67	97.1
1973	17.6	11.7	6.83	0.600	9.22	36.3	145	705	—	209	34.4	—	—
1974	3.35	0.607	0	0	7.67	4.90	182	796	825	680	—	22.1	—
1975	13.0	7.82	4.38	2.26	—	12.1	260	1,310	1,680	—	83.9	30.3	—
1976	18.6	13.0	8.67	5.69	—	51.2	241	816	585	560	143	28.9	—
1977	6.70	4.65	3.07	1.84	1.51	15.9	235	856	1,150	248	30.2	10.0	214
1978	14.3	9.35	5.22	7.80	28.0	50.9	190	962	1,770	463	—	—	—
1979	16.2	10.0	6.32	4.61	20.5	57.2	146	552	498	153	—	30.5	—
1980	—	—	—	—	—	23.0	301	1,180	911	—	—	—	—
1981	12.2	8.14	5.05	—	4.60	17.0	170	503	1,000	253	51.5	20.4	—
1982	10.2	6.09	4.37	—	—	11.9	13.1	3.32	3.44	—	—	1.44	—
1983	4.98	15.6	—	17.6	53.7	45.1	46.3	47.9	217	60.3	51.2	—	—
1984	64.9	73.3	89.8	—	56.7	61.9	49.9	36.6	49.7	53.9	51.5	27.2	—
1985	32.4	28.6	27.0	23.9	—	—	75.3	46.8	56.0	—	—	—	—
1986	52.8	65.0	77.4	90.5	84.3	73.7	60.1	40.2	53.3	51.7	47.3	46.8	61.9
1987	—	—	60.5	91.0	110	109	123	30.3	24.7	17.5	9.28	—	—
1989	—	—	—	—	—	35.2	—	—	—	—	—	—	—
1990	34.2	29.4	34.2	36.9	—	31.3	82.5	806	197	32.3	30.0	13.7	—
1991	13.5	31.2	38.4	38.5	24.0	31.7	23.4	1,050	311	13.1	12.5	19.0	134
1992	27.1	25.4	25.1	18.1	—	—	33.0	—	—	—	—	—	—
1995	15.4	—	—	—	—	—	—	—	—	—	—	—	—
1999	—	—	—	—	10.4	—	—	—	—	—	—	—	—
Average	14.3	11.7	11.8	11.2	16.1	36.3	176	710	1,020	392	63.8	19.7	212

Source: http://aochycos.ird.ne/HTMLF/ORGINT/HYDRONIG/INDEX.HTM.

Note: — = not available.

a. Riao. Country: Cameroon; basin: Niger; river: Benue; area: 27,600 square kilometers; altitude: 185 meters; latitude 9.03.00 N; longitude 13.42.00 E.

Table A2.21 A.B.N/HYDRONIGER—Hydrometry—C.I.P/Niamey, Average Monthly and Annual Flows, Station 1051500024 Buffle Noir[a]

cubic meters per second

Year	Jan.	Feb.	March	April	May	June	July	Aug.	Sept.	Oct.	Nov.	Dec.	Annual
1955	—	—	—	—	—	—	—	—	—	223	27.7	9.51	—
1956	5.09	—	—	—	—	—	49.0	186	259	131	14.8	6.74	—
1957	3.25	1.67	.967	1.60	4.45	—	—	175	—	—	30.7	9.58	—
1958	4.54	2.10	.935	1.79	4.29	12.6	64.0	94.9	206	91.6	15.8	6.54	42.1
1961	5.87	2.89	1.12	1.03	3.32	30.7	110	82.5	161	66.3	12.8	5.41	40.3
1962	2.80	—	—	—	6.80	34.4	59.3	127	190	98.1	22.5	7.87	—
1963	3.77	1.82	0.806	1.79	7.06	6.56	36.1	267	179	105	18.2	7.29	52.9
1964	3.64	1.51	—	—	5.64	17.2	82.4	102	255	89.4	28.0	9.06	—
1965	4.45	1.82	1	0.800	2.35	11.8	48.6	165	220	57.5	12.1	5.19	44.2
1966	2.48	1.03	0.419	1.39	12.5	34.0	45.0	223	283	93.3	19.8	6.03	60.2
1967	3.06	1.46	0.709	0.633	5.70	20.1	92.8	160	181	94.4	14.1	4.51	48.2
1968	2.35	1.10	0.451	3.96	4.64	10.6	73.7	198	211	58.1	11.3	3.16	48.2
1969	1.54	0.821	—	—	4.80	25.1	89.5	249	251	77.6	22.4	6.38	—
1970	2.64	0.750	0	0.266	5.41	14.7	62.6	208	235	57.5	12.0	4.32	50.3
1971	1.90	0.571	0	0.233	0.967	10.3	74.6	163	176	36.8	8.10	2.61	39.6
1972	1.00	0	0	0.533	7.25	16.1	45.0	106	83.3	79.0	11.6	4.45	29.5
1973	1.25	—	—	—	4.32	8.63	32.7	115	118	38.6	7.26	2.64	—
1974	1.00	0.071	—	0.566	5.83	3.83	41.0	97.9	142	113	—	—	—

(Table continues on the following page.)

Table A2.21 (continued)

Year	Jan.	Feb.	March	April	May	June	July	Aug.	Sept.	Oct.	Nov.	Dec.	Annual
1975	2.48	1.03	0	0	1.25	4.03	66.7	171	247	122	15.7	6.00	53.1
1976	2.93	1.31	0.225	0	3.51	12.5	83.0	180	121	114	29.9	8.67	46.4
1977	3.83	1.41	0.389	0.079	1.29	10.8	66.8	134	172	83.3	9.05	3.51	40.5
1978	1.38	0.428	0	—	6.51	27.2	46.6	134	177	80.1	27.1	6.45	—
1979	—	—	—	—	—	—	—	—	—	—	—	—	—
1980	—	0.510	—	—	—	—	—	—	—	—	—	—	—
1983	—	0.867	0.217	—	—	—	—	—	—	—	—	—	—
1984	0.301	0.109	0.039	0.461	0.971	1.01	27.1	40.1	—	—	—	—	—
1985	—	—	—	—	—	—	9.08	26.2	—	—	3.63	1.35	—
1986	—	—	—	—	1.93	5.74	23.9	80.7	79.5	54.0	6.47	2.07	—
1987	0.631	—	—	—	—	—	—	12.3	—	—	—	—	—
1989	—	—	—	—	—	—	—	—	—	—	—	2.22	—
1990	0.946	0.360	0.021	0.326	3.00	4.41	39.6	115	100	38.0	—	—	—
1994	—	—	—	—	—	—	—	—	—	—	—	—	—
1995	1.88	1.18	0.652	0.588	4.01	—	—	225	143	—	—	—	—
1997	—	—	—	—	—	—	—	—	—	—	—	—	—
Average	**2.60**	**1.08**	**0.418**	**0.891**	**4.49**	**14.7**	**57.1**	**142**	**182**	**87.0**	**16.6**	**5.48**	**45.8**

Source: http://aochycos.ird.ne/HTMLF/ORGINT/HYDRONIG/INDEX.HTM.

Note: — = not available.

a. Buffle Noir. Country: Cameroon; basin: Niger; river: Benue; area: 3,220 square kilometers; altitude: 350 meters; latitude 8.52.00 N; longitude 13.54.00 E.

Table A2.22 Specific Flow in the Upper Basins of the Niger River and Bani River

Station and year	Annual flow (l/s/km^2)	Suspension (t/yr/km^2)	Dissolved solids (t/yr/km^2)
Niger to Banankoro			
1990	7.4	—	—
1991	7.6	8.1	10.4
1992	7.4	6.6	11.7
Niger to Koulikoro			
1990	6.1	—	—
1991	6.4	7.4	8.0
1992	6.4	6.3	7.8
Bani to Douna			
1990	1.5	2.8	—
1991	1.9	3.2	2.7
1992	1.4	2.5	2.5

Source: Olivry and others 1995.

Note: — = not available.

Figure A2.5 TDS Concentrations and TDS Daily Flows for the Niger River at Banankoro (a) and the Bani River at Douna (b)

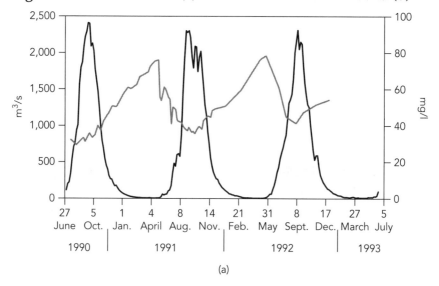

(a)

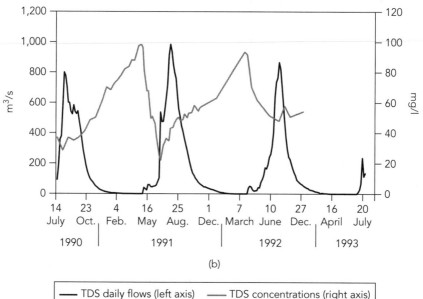

(b)

— TDS daily flows (left axis)   — TDS concentrations (right axis)

Source: Olivry and others 1998.

Figure A2.6 Average Monthly Suspended Solids Concentrations and Monthly Flows for the Niger River at Banankoro (a), at Koulikoro (b), and the Bani River at Douna (c)

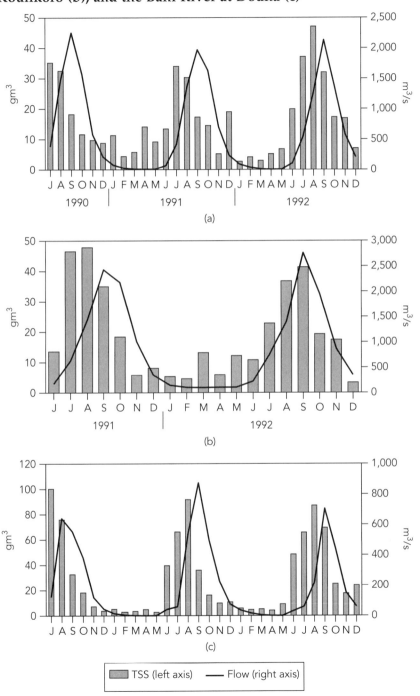

Source: Olivry and others 1998.

Figure A2.7 The Ionic Composition (average interannual ratio of each cation and anion in micro equivalent per liter) of the Niger River at Banankoro (a) and the Bani River at Douna (b)

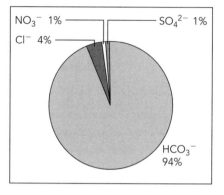
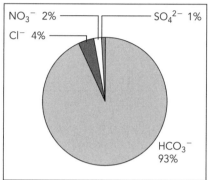
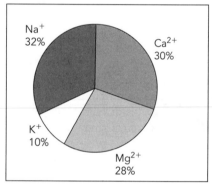
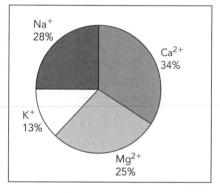

(a) Niger at Banankoro   (b) Bani at Douna

Source: Picouet 1999.

Appendix 3: Overview of Data Management

In 1985, the Hydrological Forecasting System in the Niger River Basin (HYDRONIGER) project became operational when the International Center for Hydrological Planning in Niamey was commissioned, under the direction of the Niger Basin Authority (NBA). This project provided a reception station, installation of 65 data collection platforms, a supply of data processing software (Hydrom) designed by the Office pour la Recherche Scientifique et Technique d'Outre Mer (Office for Overseas Scientific and Technical Research; ORSTOM), and training of hydrology technicians. Financed by the United Nations Development Programme (UNDP), the European Economic Community, and the Organization of Petroleum Exporting Countries, the project is managed by the World Meteorological Organization (WMO). Currently, only 15 stations are operational. The World Health Organization–Onchocerciasis Control Programme in West Africa (WHO-Oncho) program was one of the first users of this technology in the region; data collection platforms were installed in the northern part of Côte d'Ivoire, in southern Mali, and in Guinea, among others, along with the Argos direct reception stations in the offices of the WHO-Oncho program in Bobo-Dioulasso, Bamako, and Odienne.

The primary hydrology research work on the Niger River to the Nigerian border is the ORSTOM Hydrological Monograph No. 8 in two volumes, published in 1986 under the direction of Y. Brunet-Moret, with the assistance of P. Chaperon, J. P. Lamagat, and M. Molinier. The monograph covers data up to 1979. For Benin, Chad, and Cameroon, ORSTOM monographs were also used (Olivry 1986; Le Barbe 1990). For Nigeria, information has been gathered from various sources; a comprehensive overview is not currently available.

In 1992, for West and Central Africa, country reports were published covering the management of hydroclimatic data collection and databases by agencies for meteorology, groundwater, and hydrology. This 23-volume *Water Assessment Project* was a significant and informative undertaking, despite some gaps in coverage and quality. Meteorological data from synoptic observation stations are generally very well monitored.[11] Rainfall stations are well maintained, although dissemination of information could be improved, specifically in providing farmers with relevant meteorological information in a timely manner. Hydrologic data on water levels are generally insufficient. Limnimetric stations need to be standardized with gauges to measure hydrologic parameters. The national water departments are responsible for hydrometric network maintenance, for which they often receive financing from bilateral cooperation agencies. Generally, the practice is to attain the standards set by the "Operational Hydrology Practices" of the WMO in terms of network density, although these may not be the best adapted to the countries of Sub-Saharan Africa.

In 1993, the WMO promoted a global observation system for the hydrological cycle (World Hydrological Cycle Observing System; WHYCOS), using monitoring reference stations (hydrological observatories) with real time or near real time data transmission (Rodda and others 1993). The pilot Afrique de l'Ouest et Centrale (West and Central Africa)–Hydrological Cycle Observing System (AOC-HYCOS) project was launched in 1996 in Ouagadougou, Burkina Faso, with French funding and was renewed in January 2000. The focus of this project was to allow national hydrological departments to post and share observations on a central Web site created by the Ecole Inter-Etats d'Ingénieurs de l'Equipement Rural (Interstate Institute for Rural Engineering) in Ouagadougou. Subsequently, the project was headquartered in Niamey, Niger, at the NBA's Centre Inter-Etats de Prévision (Interstate Forecasting Center; CIP), in association with the Centre Régional de Formation et d'Application de la Météorologie et de l'Hydrologie Opérationnelle (Regional Center for Training and Application of Meteorology and Operational Hydrology; AGRHYMET).[12]

There are other national or subregional monitoring activities in the Basin. The Gestion hydro-écologique du Niger supérieur Mali-Guinée (GHENIS) project (financed by the Netherlands) supports data collection platforms and remote satellite transmission, including hydrometric sensors and water quality measurement sensors (for example, conductivity, turbidity, and temperature) in the Upper Niger River Basin (Mali and Guinea) from Bamako. In Nigeria, the National Water Resources Master Plan lays out hydrological areas (HA). For the Niger River Basin, there are the Northwest Region HA-I, the Central West Region HA-II, the Central East Regions HA-III/IV, and the South East Regions HA-V/VII, in which various river basin development authorities manage most of the water measuring stations. The Federal Department for Navigable Waterways manages its own stations and the National Electric Power Authority manages dam sites. In Cameroon, hydrologists from the Institut de Recherches Géologiques et Minières (Institute for Geological and Mining Research) in charge of a national network no longer have access to data collected by the National Electricity Company, which manages its own network since it was privatized. Other methods also exist, including user-friendly on-site data transmission by satellite from manual test units. This system is already in use for meteorology and by the United Nations Food and Agriculture Organization in village markets to determine daily commodity prices. There is other progress in hydrology, such as flow measurement by the Acoustic Doppler Current Profiler, which is a simple exploration of sections through a Doppler monitoring radar (for example, small sections of a shallow river).

Appendix 4: Glossary

Acacia	Genus *Robinia*, any of various spiny trees or shrubs of the genus Acacia, similar to the locust.
Average annual concentrations	The annual ratio between TSS flow and water flow, and the rate of specific transport (Ts).
Base flow	The normal sustained low flow of the river, originating entirely from groundwater discharging to the river.
Boreal	Of or relating to the north; northern, of or concerning the north wind.
Bourgou grass	*Echinochloa stagnina*. A tropical aquatic perennial growing along rivers and in lakes and lagoons in water up to 3 meters deep, the stems of which root at the nodes, and produce excellent regrowth for grazing during the dry season.
Continental Terminus	Composed of claylike sandstone, sand, and clays. A continuous stratum aquifer with good water quality located within the Niger Basin.
Endoreism	A water body that remains isolated and cut off without any geomorphic means to connect to the water network.
Erg	Also called "Sand Sea." In a desert region, area of large accumulation of sand.
Ferralitic and ferruginous soils	Relating to or containing iron, characterized by high sand content.
Fouta Djallon	The mountainous massif known as the West African headwaters.
Guinée Forestière	Includes the southern high plateaus of the Fouta Djallon Massif.
Harmattan	A dusty and hot wind from the Sahara that blows toward the western coast of Africa during the winter (December–May).
Haute Guinée	Includes the northern and eastern highlands of the Fouta Djallon Massif.
Hydrography	Refers to water and drainage features and description of physical properties of the waters of a region.
Hydromorphic soils	Characterized by the temporary or permanent presence of a surface water table.
Interannual flow	Average yearly flows for a period of years.
Isohumic soils	Sahelian soils poor in organic matter with a deep homogenous sandy profile.
Isohyets	Relating to or indicating equal rainfall.

Lateritic butte	A steep-sided, flat-topped hill formed by erosion of flat laying strata, where remnants of a resistant layer (in this case, lateritic soil, see below) protect the softer rocks underneath.
Lateritic soils	A red residual soil in humid tropical and subtropical regions that is leached of soluble minerals, aluminum hydroxides, and silica, but that still contains concentrations of iron oxides and iron hydroxides. Hardens on exposure to the atmosphere.
Limnologic stations	Monitoring stations for gathering physical, chemical, meteorological, and biological data on the conditions in fresh waters.
Lithosol	A type of azonal soil having no clear expression of soil morphology consisting of freshly and imperfectly weathered mass.
Monsoon	A humid and wet wind blowing from the southwest in the summer (June–November) and responsible for the rainy season in West Africa.
Niger Bend	Refers to the major change in the Niger River flow direction from south-southwest to east-southeast, downstream from the Inland Delta in Mali, upstream from the city of Gao.
Office du Niger	Refers to the large irrigation area in Mali, and the Malian government agency, which manages the irrigation schemes.
Reach	An extended portion of a waterway.
Synoptic stations	Monitoring stations that gather meteorological data over large areas at a specified instant in time, for the purpose of projecting the data into the future, that is, to give weather forecasts.
TDS	Total dissolved solids. The standard measure of minerals dissolved in water. Used to evaluate water quality.
TSS	Total suspended solids. The total of all settleable and nonsettleable solids in a sample of wastewater, measured in milligrams per liter.
Tussocky grass	*Sporobolus indicus*. A compact tuft especially of grass or sedge; also an area of raised solid ground in a marsh or bog that is bound together by roots of low vegetation.
Vertisol	Tropical brown soil or tropical black clay characterized by at least 30 percent clay.
Wadi	A desert watercourse that is usually dry and contains water only occasionally, after heavy rainfall.

Endnotes

1. Benin, Burkina Faso, Cameroon, Chad, Côte d'Ivoire, Guinea, Mali, Niger, and Nigeria.

2. Based on 2002 gross national income (GNI) per capita (World Bank 2004b).

3. The HDI ranges from 1 to 177, with the latter being the lowest. Eight of the nine Basin countries fall into the "Low Human Development Category," ranked at 142 and below (UNDP 2004).

4. In the era when Bismarck was the arbiter of Europe and Germany wanted to create "its" colonies, sections of Africa were defined for each colonial power: Cameroon for Germany, Nigeria for Great Britian, and Nigerian Sudan and Oubangui-Chari for France.

5. This includes projects that involve or have impacts on navigation, agriculture, hydropower, industry, water quality, and flora and fauna.

6. For example, HYDRONIGER, directed by the World Meteorological Organization (WMO), had been financed by a number of donor institutions and bilaterals.

7. This commitment was further reinforced by a cooperative framework confirmed by the attending donors to support this process.

8. Nigeria extends across four climatic zones: semiarid, tropical pure, transitional tropical, and equatorial in the Lower Niger Delta.

9. For further discussion of benefits, see Sadoff and Grey (2002).

10. The principle of subsidiarity is regularly used in the European Union context, and suggests that implementation of a particular policy is delegated to the lowest appropriate level. In the river basin context, the principle has been used to encourage decisions being similarly managed at the lowest appropriate level, whether at the local, national, or subbasin level.

11. The *Water Assessment Project* was prepared by Mott MacDonald International, BCEOM, SOGREAH, and the French institute, ORSTOM, with financial support from the World Bank, UNDP, the African Development Bank, and the French Cooperation Ministry.

12. AGRHYMET is a special institution of the Comité Permanent Inter-Etats de Lutte contre la Sécheresse au Sahel (Permanent Interstate Committee for Drought Control in the Sahel).

Bibliography

Archambault, J. 1960. *Les eaux souterraines de l'Afrique Occidentale*. A. O. F. Hydraulic Service.

Auvray, J. 1960. *Monographie du Niger. B- La cuvette lacustre*. Vol. 1. ORSTOM, Bamako, Mali, 12–45.

Bamba, F., M. Diabate, G. Mahé, and M. Diarra. 1996. "Rainfall and Runoff Decrease of Five River Basins of the Tropical Upstream Part of the Niger River over the Period 1951–1989." In *Global Hydrological Change*, ed. L.A. Roald. XXIst General Assembly, European Geological Association, The Hague, The Netherlands, May 6–10.

Bamba, F., G. Mahé, J. P. Bricquet, and J. C. Olivry. 1996a. "Changements climatiques et variabilité des ressources en eau des bassins du Haut Niger et de la Cuvette Lacustre." In *Réseaux hydrométriques, réseaux télématiques, réseaux scientifiques: nouveaux visages de l'Hydrologie Régionale en Afrique*. XIIèmes Journées Hydrologiques de l'ORSTOM, October 10–11, Montpellier, France.

———. 1996b. "Changements climatiques récents et modifications du fleuve Niger à Koulikoro (Mali)." In *L'Hydrologie Tropicale. Géoscience et outil pour le développement*. IAHS Series of Proceedings and Reports 238: 157–66.

Barral, J. P. 1997. "Evolution des sols sous irrigation à l'Office du Niger. La gestion de l'eau sur les périmètres irrigués à l'Office du Niger." Activity Report 96-97. Institut d'économie rurale. Pôle systèmes irrigués. CORAF, FAC 93014800, Mali.

Barral, J. P., and M. K. Dicko. 1996. "La dégradation des sols à l'Office du Niger." Activity Report 96-97. Institut d'économie rurale. Pôle systèmes irrigués. CORAF, FAC 93014800, Mali.

Beets, C. 1988. "Field Studies for Flood and Erosion in the Niger Delta." Internal report IGST, Rivers State University of Science and Technology. Port Harcourt, Nigeria.

Blanck, J. P., and G. Lutz. 1990. "Rapport de mission au Mali: Évaluation des ressources en terres et en eaux du delta central du Niger, du 26 novembre au 14 décembre 1989." Strasbourg, France: ULP.

Blanck, J. P., and J. L. F. Tricart. 1990. "Some Effects of Neotectonics on the Landforms in the Region of the Central Delta of Niger River (Mali)." *Comptes-Rendus de l'Académie des Sciences* 310 (2): 309–13.

Bleich, K. E., L. Herrmann, K. H. Papenfuss, and K. Stahr. 1994. "Dust Influx into the Soils of the Sahélian Zone in Niger: Its Composition and Identification." In *Wind Erosion in West Africa: The Problem and Its Control. Proceedings of the*

International Symposium, University of Hohenheim, Germany, 5–7 December 1994, ed. B. Buerkert and others. Weikersheim, Germany: Margraf Verlag.

Boeglin, J.-L., and J. L. Probst. 1996. "Transports fluviaux de matières dissoutes et particulaires sur un bassin versant en region tropicale: Le bassin amont du Niger au cours de la période 1990–1993." *Sciences Géologiques Bulletin* 49 (1): 25–45.

Boeglin, J.-L., and Y. Tardy. 1997. "Erosion chimique et mécanique sur le bassin amont du Niger (Guinée, Mali): Découpage d'hydrogramme en quatre écoulements." *Comptes-Rendus de l'Académie des Sciences. Série 2a: Sciences de la Terre et des Planètes* 325 (2): 125–31.

Boeglin, J.-L., J. Mortatti, and Y. Tardy. 1997. "Erosion chimique et mécanique sur le bassin amont du Niger (Guinée, Mali): Bilan géochimique de l'altération en milieu tropical." *Comptes-Rendus de l'Académie des Sciences. Série 2a: Sciences de la Terre et des Planètes* 325 (3): 185–91.

Bonnefoy, A. 1998. "Impact des intrants agricoles sur la qualité des eaux en zone cotonnière du sud Mali." Bamako, Mali: Institut Universitaire Professionnalisé Environnement, Technologie; and ORSTOM.

Bonneval, P., M. Kuper, and J. P. Tonneau. 2002. *L'Office du Niger, grenier à riz du Mali. Succès économiques, transitions culturelles et politiques de développement.* Montpellier, France/Paris: Cirad/Karthala.

Bricquet, J. P., G. Mahé, F. Bamba, and J. C. Olivry. 1995. "Changements climatiques récents et modifications du fleuve Niger à Koulikoro (Mali)." In *L'Hydrologie Tropicale. Géoscience et outil pour le développement* (Colloque J. Rodier, May 2–3). IAHS Series of Proceedings and Reports 238: 157–66.

Bricquet, J. P., F. Bamba, G. Mahé, M. Toure, and J. C. Olivry. 1997. "Evolution récente des ressources en eau de l'Afrique atlantique." *Revue des Sciences de l'eau* 3: 321–37.

Brunet-Moret, Y., P. Chaperon, J. P. Lamagat, and M. Molinier. 1986. *Monographie hydrologique du fleuve Niger. Tome I: Niger supérieur; Tome II: Cuvette Lacustre et Niger moyen.* Colloque Monographies Hydrologiques 8. Paris: ORSTOM.

Carbonnel, J. P., and P. Hubert. 1992. "Pluviométrie en Afrique de l'Ouest soudano-sahélienne: remise en cause de la stationnarité des séries." In *L'aridité: Une contrainte au développement*, ed. E. Le Floc'h and others. Paris: ORSTOM.

Censier, C., J. C. Olivry, and J. P. Bricquet. 1995. "Les apports détritiques terrigènes dans la cuvette lacustre du Niger entre Mopti et Kona (Mali)." In *Grands bassins fluviaux péri-atlantiques: Congo, Niger, Amazone*, ed. J. C. Olivry and J. Boulègue. Proc. PEGI/INSU/CNRS Symposium, November 22–24, 1993. Paris: ORSTOM.

Chamard, P., M. F. Courel, and M. Adesir-Schilling. 1997. "L'inondation des plaines du delta intérieur du Niger (Mali). Tentatives de contrôle: la réalité et les risques." *Sécheresse* 8 (3): 151–56.

CIDA (Canadian International Development Agency). 2004a. Etude multisectorielle nationale, "Evaluation des opportunités et contraintes au développement dans la portion beninoise du Bassin du Fleuve Niger."

———. 2004b. Etude multisectorielle nationale, "Evaluation des opportunités et contraintes au développement dans la portion burkinabaise du Bassin du Fleuve Niger."

———. 2004c. Etude multisectorielle nationale, "Evaluation des opportunités et contraintes au développement dans la portion ivoirienne du Bassin du Fleuve Niger."

———. 2004d. Etude multisectorielle nationale, "Evaluation des opportunités et contraintes au développement dans la portion malienne du Bassin du Fleuve Niger."

———. 2004e. National multi-sector study on "Assessment of the Opportunities and Constraints to the Development of the Niger's Portion of the River Niger Basin."

———. 2004f. Etude multisectorielle nationale, "Evaluation des opportunités et contraintes au développement dans la portion nigérienne du Bassin du Fleuve Niger."

Collignon, B. 1994. "Impact des activités humaines sur les ressources en eau souterraine en Afrique sahélienne et soudanienne." In *Enregistreurs et indicateurs de l'évolution de l'environnement en zone tropicale*, ed. K. Maire and others. Talence, France: University of Bordeaux Press.

Cunge, J. A., F. M. Holly, and A. Verwey. 1980. *Practical Aspects of Computational River Hydraulics*, Vol. 3 of *Monographs and Surveys in Water Resources Engineering*. London: Pitman Publishing Ltd.

Dante, Y. N. T. 1995. "Description des acteurs concernés par le devenir du fleuve Niger au Mali sur les problèmes d'environnement." Project RAF/95/G45/GEF/PDF. Preparatory Assistance on the Niger River. UN-DADSG-UNDP.

Dejoux, C. 1988. "La pollution des eaux continentales africaines: expérience acquise, situation actuelle et perspectives." *Travaux et Documents* 213. Paris: ORSTOM.

Diarra, A., and A. Soumaguel. 1997. "Influence of the Selingue Dam on the Hydrological System of the Niger River." In *Sustainability of Water Resources under Increasing Uncertainty. Proceeding of the Rabat Symposium, April 1997*. IAHS Series of Proceedings and Reports 240: 277–86.

Favreau, G., and C. Leduc. 1998. "Fluctuations à long terme de la nappe phréatique du Continental Terminal près de Niamey (Niger) entre 1956 et 1997." In *Water Resources Variability in Africa during the XXth Century. Proceedings of the Abidjan 1998 Conference*. IAHS Series of Proceedings and Reports 252: 253–58.

Fontes, J. C., J. N. Andrews, W. M. Edmunds, A. Guerre, and Y. Travi. 1991. "Paleorecharge by the Niger River (Mali) Deduced from Groundwater Geochemistry." *Water Resources Research* 27 (2): 199–214.

Gallaire, R. 1995. "Données sur les transports du Niger moyen entre Kandadji et Niamey." In *Grands bassins fluviaux péri-atlantiques: Congo, Niger, Amazone*, ed. J. C. Olivry and J. Boulègue. Proc. PEGI/INSU/CNRS Symposium, November 22–24, 1993. Colloques et Séminaires ORSTOM, 267–80.

Gallais, J. 1967. *Le delta intérieur du Niger et ses bordures. Etude morphologique.* Memoires and Documents: Center for Cartographic and Geographic Research and Documentation. Paris: CNRS.

———. 1979. "Etude morphologique au niveau du lac Débo." In *Détermination des causes des anomalies de la crue du Niger.* Paris: ORSTOM.

Gerbe, A. 1994. *Etude de l'ensablement de la vallée du fleuve dans la boucle du Niger au Mali.* Montpellier, France: IARE.

Godana, B. A. 1985. *Africa's Shared Water Resources. Legal and Institutional Aspects of the Nile, Niger and Senegal River Systems.* London: Frances Pinter Publishers; Boulder, Colorado: Lynne Rienne Publishers.

Gourcy, L. 1993. "Note sur les mesures isotopiques du Niger à Banankoro et du Bani dans la région de Douna-San." In *Les ressources en eau au Sahel (Études hydrogéologiques et hydrologiques en Afrique de l'ouest par les techniques isotopiques) (Comptes rendus des études effectuées dans le cadre du projet RAF/8/012: Hydrologie isotopique dans les pays du Sahel).* Vienna: IAEA TECDOC Series 721.

———. 1994. "Fonctionnement hydrogéochimique de la cuvette lacustre du fleuve Niger (Mali)." Doctoral thesis, University of Paris XI, Orsay.

Grey, D., and C. Sadoff. 2002. "Water Resources and Poverty in Africa: Essential Economic and Political Reponses." Discussion paper presented at the African Ministerial Conference on Water, Abuja, Nigeria, 29–30 April.

Grouzis, M. 1992. "Germination et établissement des plantes annuelles sahéliennes." In *L'aridité: Une contrainte au développement,* ed. E. Le Floc'h and others. Paris: ORSTOM.

Guerre, A., and J. F. Aranyossy. 1989. Synthèse des ressources en eau du Mali. Résultats et interprétation des analyses isotopiques. Regional African Project Report RAF/8/012. Vienna: IAEA TECDOC Series 721.

Guiguen, N. 1985. "Etudes hydrologiques complémentaires de la cuvette lacustre du Niger." HYDRONIGER Project. Bamako, Mali: ORSTOM.

Hassane, A., M. Kuper, and D. Orange. 2000. "Influence des aménagements hydrauliques et hydroagricoles du Niger supérieur sur l'onde de la crue du delta intérieur du Niger au Mali." *Sud Sciences et Technologies* 5: 16–31.

Hydroconsult. 1995. "Hydrologie pour la réhabilitation du barrage de Sélingué (Mali)." International Hydroconsult Contract Operator/ORSTOM 9512.

———. 1996a. "Mise à jour de l'hydrologie pour la réhabilitation du barrage de Sélingué." Two volumes. ORSTOM-EDF, Bamako, Mali.

———. 1996b. "Etude de factibilité et d'impact du barrage de Tossaye—Hydrologie." DNHE-GIE ORSTOM/EDF.

———. 1996c. "Etude de factibilité et d'impact du barrage de Tossaye—Calcul des courbes de remous. Note méthodologique." DNHE-GIE ORSTOM/EDF.

Iwaco B. V., and Delft Hydraulics. 1996. "Projet pilote Guinée-Mali de création d'un système intégré de suivi hydro-écologique du bassin de Niger supérieur. Rapport final." Ministry of Energy and Environment and Ministry of Agriculture, Livestock and Forests, Conakry, Guinea; and Ministry of Rural

Development and Environment and Ministry of Energy and Water, Bamako, Mali. Two volumes.

Jaccon, G. 1968. "La crue exceptionnelle du Niger en 1967." *Cahiers ORSTOM, Hydrology Series* V (1): 15–53.

JICA (Japan International Cooperation Agency). 1995. *Master Plan of the Federal Ministry of Water Resources and Rural Development. Study on the National Water Resources Master Plan* (NWRMP). Five volumes. Abuja, Nigeria.

Joignerez, A., and N. Guiguen. 1992. "Evaluation des ressources en eau non pérennes du Mali." Final ORSTOM Report, Paris.

Kuper, M., A. Hassane, D. Orange, A. Chohin-Kuper, and M. Sow. 2002. "Régulation, utilisation et partage des eaux du fleuve Niger: L'impact de la gestion des aménagements hydrauliques sur l'Office du Niger et le delta intérieur du Niger au Mali." Colloques et séminaires. Bamako, Mali: IRD.

Lamagat, J. P., and M. Molinier. 1983. *Etude des anomalies des crues du Niger.* Travaux et Documents No. 161. Paris: ORSTOM.

Lamagat, J. P., S. Sambou, and J. Albergel. 1996. "Analyse statistique de l'écoulement d'un fleuve dans une plaine d'inondation: application aux côtes maximales du fleuve Niger dans la cuvette lacustre." In *L'hydrologie tropicale: Géoscience et outil pour le développement.* IAHS Series of Proceedings and Reports 238: 367–79.

Lapie. 1829. *Atlas universal de géographie ancienne et moderne.*

Le Barbe, L., G. Ale, B. Millet, H. Texier, Y. Borel, and R. Gualde. 1990. *Les resources en eaux superficielles de la république du Bénin. Monographie hydrologique.* Paris: ORSTOM.

Le Barbe, L., and T. Lebel. 1997. "Rainfall Climatology of the HAPEX-Sahel Region during the Years 1950–1990." *Journal of Hydrology* 188 (1): 43–73.

Le Barbe, L., G. Ale, B. Millet, H. Texier, Y. Borel, and R. Gualde. 1990. *Les ressources en eaux superficielles de la république du Bénin. Monographie hydrologique.* Paris: ORSTOM.

Leroux, M. 1996. *La dynamique du temps et du climat.* Paris: Masson.

L'Hôte, Y., and G. Mahé. 1996. *Afrique de l'Ouest et Centrale. Carte des précipitations moyennes annuelles (période 1951–1989).* Paris: ORSTOM.

LOTTI and SOFRELEC. 1975. *Barrage de Sélingué sur le Sankarani, recommandations sur les problèmes de gestion et de fonctionnement.* Bamako, Mali.

Mahé, G. 1993. *Les écoulements fluviaux sur la façade Atlantique de l'Afrique. Etudes des éléments du bilan hydrique et variabilité interannuelle. Analyse de situations hydroclimatiques moyennes et extrêmes.* College Etudes et Thèses. Paris: ORSTOM.

Mahé, G., and J. C. Olivry. 1991. "Les changements climatiques et variations des écoulements en Afrique occidentale et centrale, du mensuel à l'interrannuel." In *Hydrology for the Water Management of Large Rivers Basins. Proceeding of the Vienna Symposium, August 1991.* IAHS Series of Proceedings and Reports 201: 163–72.

Mahé, G., and J. C. Olivry. 1995. "Variation des précipitations et dés écoulements en Afrique de l'Ouest et Centrale." *Sécheresse* 6 (1): 109–17.

Mahé, G., J. P. Bricquet, A. Soumaguel, F. Bamba, M. Diabate, T. Henry Des Tureaux, C. Konde, J. F. Leroux, A. Mahieux, J. C. Olivry, D. Orange, and

C. Picouet. 1997. "Bilan hydrologique du Niger à Koulikoro depuis le début du siècle." In *LOC*, ed. M. Mikos. *Proceedings of Oral Presentations - FRIEND '97*. Acta Hydrotechnica 15/18, October 1-4, 1997, Ljubljana, Slovenia.

Mahé, G., R. Dessouassi, B. Cissoko, and J. C. Olivry. 1998. "Comparaison des fluctuations interannuelles de piézométrie, précipitation et de débit sur le bassin versant du Bani à Douna au Mali." In *Water Resources Variability in Africa during the XXth Century Proceedings of the Abidjan 1998 Conference*. IAHS Series of Proceedings and Reports 252: 289–96.

Maiga, H. A. 1998. "Effets des sécheresses et étiages dans le bassin moyen du fleuve Niger au Mali." In *Water Resources Variability in Africa during the XXth Century. Proceedings of the Abidjan 1998 Conference*. IAHS Series of Proceedings and Reports 252: 437–43.

Maley, J. 1982. "Dust, Clouds, Rain Types, and Climatic Variations in Tropical North Africa." *Quaternary Research* 18: 1–16.

Mali, DNHE (Direction Nationale de l'Hydraulique et de l'Energie). ORSTOM/DNHE databank.

Marieu, B., M. Kuper, and A. Mahieux. 2000. "Delta intérieur du Niger—Synthèse des résultats hydrologiques acquis en 1998–1999." Montpellier, France: MSE-IRD.

Mariko, A., G. Mahé, D. Orange, E. Servat, and A. Amani. 2000. "Utilisation de la télédétection NOAA/AVHRR et analyse hydrologique de l'inondation dans le delta intérieur du Niger (Mali)." *Séminaire international sur la gestion intégrée des ressources naturelles en zones inondables tropicales (GIRN-ZIT)*, Bamako, Mali, June 20–23.

McCarthy, T. S. 1993. "The Great Inland Deltas of Africa." *Journal of African Earth Sciences* 19 (3): 275–91.

Meybeck, M. 1984. Les fleuves et le cycle géochimique des éléments. Thèse Doctoral Etat. Sciences. University of Paris, Paul and Marie Curie, Paris XI.

Milliman, J. D., and P. M. Syvitski. 1992. "Geomorphic/Tectonic Control of Sediment Discharges to the Ocean: The Importance of Small Mountainous Rivers." *Journal of Geology* 100: 524–44.

Mott Macdonald Int., BCEOM, SOGREAH, ORSTOM. 1992a. *Evaluation de l'Afrique Sub-Saharienne—Pays de l'Afrique de l'Ouest. Rapport de pays: Guinée.* Report prepared for World Bank, UNDP, ADB, MFC.

———. 1992b. *Evaluation de l'Afrique Sub-Saharienne—Pays de l'Afrique de l'Ouest. Rapport de pays: Mali.* Report prepared for World Bank, UNDP, ADB, MFC.

———. 1992c. *Evaluation de l'Afrique Sub-Saharienne—Pays de l'Afrique de l'Ouest. Rapport de pays: Niger.* Report prepared for World Bank, UNDP, ADB, MFC.

———. 1992d. *Evaluation de l'Afrique Sub-Saharienne—Pays de l'Afrique de l'Ouest. Rapport de pays: Nigéria.* Report prepared for World Bank, UNDP, ADB, MFC.

Mudry, J., and Y. Travi. 1994. "Sécheresse sahélienne et action anthropique, deux facteurs conjugués de dégradation des ressources en eau de l'Afrique de l'Ouest." In *Enregistreurs et indicateurs de l'évolution de l'environnement en zone tropicale*, ed. K. Maire and others. Talence, France: University of Bordeaux Press.

Nouvelot, J. F. 1969. "Mesure et étude des transports solides en suspension au Cameroun." *Cahiers ORSTOM, Série Hydrologie* 6 (4): 43–85.

Olivry, J. C. 1978. "Transports solides en suspension au Cameroun." *Cahiers de l'ONAREST* 1 (1): 47–60.

———. 1986. *Fleuves et rivières du Cameroun*. Monographies Hydrologiques. Paris: ORSTOM.

———. 1987. "Les conséquences durables de la sécheresse actuelle sur l'écoulement du fleuve Sénégal et l'hypersalinisation de la basse Casamance." In *The Influence of Climate Change and Climate Variability on the Hydrologic Regime and Water Resources*. Proceedings of a symposium held during the XIX Assembly of the International Union of Geodesy and Geophysics at Vancouver, August 1987. IAHS Series of Proceedings and Reports 168: 501–12.

———. 1997. "Long Term Effects of Rain Shortage: The Ill Rivers of Western and Central Africa." FRIEND (Flow Regimes from International Experimental and Network Data) Report No. 3. Paris: UNESCO.

———. 2002. "Synthèse de connaissance hydrologique et potential en resources en eau." Unpublished paper.

Olivry, J. C., J. P. Bricquet, and G. Mahé. 1993. "Vers un appauvrissement durable des ressources en eau de l'Afrique humide?" In *Hydrology of Warm Humid Regions, Proceeding of Yokohama Symposium, July 1993*. IAHS Series of Proceedings and Reports 216: 67–78.

———. 1995. "Fonctionnement hydrologique de la Cuvette Lacustre du Niger et essai de modélisation de l'inondation du Delta intérieur." In *Grands bassins fluviaux péri-atlantiques: Congo, Niger, Amazone*, ed. J. C. Olivry and J. Boulègue. Proceedings of the PEGI/INSU/CNRS Symposium, November 22–24, 1993. Paris: ORSTOM.

———. 1998. "Variabilité de la puissance des crues des grands cours d'Afrique intertopicale et incidence de la baisse des écoulements de base au cours des deux derniers décennies." In *Water Resources Variability in Africa during the XXth Century*. Proceedings of the Abidjan 1998 Conference. IAHS Series of Proceedings and Reports 252: 189–97.

Olivry, J. C., J. P. Bricquet, F. Bamba, and M. Diarra. 1995. "Le régime hydrologique du Niger supérieur et le déficit des deux dernières décennies." In *Grands bassins fluviaux péri-atlantiques: Congo, Niger, Amazone*, ed. J. C. Olivry and J. Boulègue. Proceedings of the PEGI/INSU/CNRS Symposium, November 22–24, 1993. Paris: ORSTOM.

Olivry, J. C., C. Picouet, D. Orange, J. P. Droux, J. P. Bricquet, A. Laraque, and J. M. Fritsch. 1996. "Transport particulaire dans le delta central du Niger (Bilan de 4 années d'observations)." Presentation at Gip-Hydrosystem Symposium, Toulouse, France.

Orange, D. 1992. "Hydroclimatologie du Fouta Djalon et dynamique actuelle d'un vieux paysage latéritique." *Sciences Géologiques Mémoire* 93. Strasbourg, France.

Pardé, M. 1933. *Fleuves et rivières*. Paris: Librairie Armand Colin.

Picouet, C. 1999. "Géodynamique d'un hydrosystème tropical peu anthropisé: Le bassin supérieur du Niger et son delta intérieur." Doctoral thesis, University of Montpellier, France.

PIRT (Project d'Inventaire des Ressources Terrestres). 1983.

Poncet, Y. 1994. "Le milieu du delta central." In *La pêche dans le delta central du Niger: Approche pluridisciplinaire d'un système de production halieutique*, ed. J. Quensière. Paris: ORSTOM, Karthala.

Pouyaud, B. 1986. "Contribution à l'évaluation de l'évaporation de nappes d'eau libre en climat tropical sec." Coll. Etude et Thèse. Paris: ORSTOM.

Quensière, J., ed. 1994. *La pêche dans le delta central du Niger: Approche pluridisciplinaire d'un système de production halieutique*. Paris: ORSTOM, Karthala.

Rey, H., B. Kassibo, and M. Salamanta. 1994. "Pirogues et constructeurs: Approche d'une activité informelle." In *La pêche dans le delta central du Niger: Approche pluridisciplinaire d'un système de production halieutique*, ed. J. Quensière. Vol. 1: 311–21. Paris: ORSTOM, Karthala.

Roche, M. F. 1963. *Hydrologie de surface*. Paris: Gauthiers-Villais.

Rodda, J. C., S. A. Pieyns, N. S. Sehmi, and G. Matthews. 1993. "Towards a World Hydrological Cycle Observing System." *Hydrological Sciences Journal* 38 (5).

Rodier, J. A. 1964. *Régimes Hydrologiques de l'Afrique noire à l'ouest du Congo*. Memoires ORSTOM 6. Paris: ORSTOM.

———. 1996. *Analyse de l'eau: eaux naturelles, eaux résiduaires, eau de mer, chimie, physico-chimie, bactériologie, biologie*. 6th ed. Paris: Dunod.

Sadoff, C., and D. Grey. 2002. "Beyond the River, The Benefits of Cooperation on International Rivers." *Water Policy* 4: 389–403.

Sadoff, C., and D. Grey. Forthcoming. "A Continuum for Securing and Sharing Benefits." *Water International*.

Sangaré, S. 2001. "Bilan Hydrologique du Niger en Guinée." *Actes du Colloque FRIEND AOC*, Capetown, March 2002.

Sanyu and others. 1995. "Study on the National Water Resources Master Plan (NWRMP)." JICA (Japan International Cooperation Agency), Tokyo.

Shiklomanov, I. A. 1998. *World Water Resources—A New Appraisal and Assessment for the 21st Century*. UNESCO IHP Nonserial Publications in Hydrology. Paris: UNESCO.

Singh, J., D. Moffat, and O. Linden. 1995. *Defining an Environmental Development Strategy for the Niger Delta*. 2 volumes. Washington, DC: Industry and Energy Operations Division, West and Central Africa Department, World Bank.

Sircoulon, J. 1976. "Les données hydropluviométriques de la sécheresse récente en Afrique Intertropicale. Comparaison avec les sécheresses 1913 et 1940. *Série Hydrologie—Cahiers ORSTOM* 13 (2): 75–174.

SOGREAH (Société Grenobloise d'Etudes et d'Applications Hydrauliques). 1992. "Barrage de Markala: Consignes générales d'exploitations et d'entretien." Grenoble, France.

SOGREAH, BCEOM, and BETICO. 1999. *Etude du schéma directeur d'aménagement pour la zone de l'Office du Niger.* Main Report. Ségou, Mali.

SONEL (Société Nationale d'Electricité de Cameroun). 1983. *Atlas du Potentiel Hydroélectrique du Cameroun.* National Electricity Company of Cameroon.

Stahr, K., L. Herrmann, and R. Jahn. 1994. "Long-Distance Dust Transport in the Sudan-Sahelian Zone and the Consequences for Soil Properties." In *Wind Erosion in West Africa: The Problem and Its Control. Proceedings of the International Symposium University of Hohenheim, Germany, 5–7 December 1994*, ed. B. Buerkert and others. Weikersheim, Germany: Margraf Verlag.

Sutcliffe, J. V., and Y. P. Parks. 1989. "Comparative Water Balances of Selected African Wetlands." *Hydrological Sciences Journal* 34: 49–62.

Toupet, C. 1976. "L'évolution du climat en Mauritanie du Moyen Age jusqu'à nos jours." In *La désertification au sud du Sahara.* Dakar, Senegal: Nouvelles Editions Africaines, 56–63.

UNDP (United Nations Development Program). 2004. *Human Development Report.* New York: United Nations. http://hdr.undp.org/reports/global/2004.

World Bank. 1986. "Sols et eaux, acquis et perspectives de la recherche agronomique française en zone intertropicale." Actes du Séminaire tenu à la Banque Mondiale May 15–16. Paris: World Bank.

———. 2004a. Etude multisectorielle nationale, "Evaluation des opportunités et contraintes au développement dans la portion camerounaise du Bassin du Fleuve Niger." Washington, DC: World Bank.

———. 2004b. *World Development Indicators.* Washington, DC: World Bank.

Yaya, I. 1995. *Etude pour l'organisation d'un Colloque sur la sauvegarde du Fleuve Niger. Rapport général introductif.* Niamey, Niger: ABN.

Zoska, J. R. 1985. "The Water Quality and Hydrobiology of the Niger." In *Niger and Its Neighbors: Environmental History and Hydrobiology, Human Use and Health Hazards of the Major West African Rivers,* ed. A.T. Grove. Boston: A. A. Balkema.

Index

Bani Watershed
   hydrography, 13–14
   hydrology, 34–36
   map, 72*m*
   TDS, 53–54, 55*f*, 124*f*
   TSS, 48–50, 49*f*, 50*t*, 55*f*, 125*f*
Benin, 2–3*t*, 4, 46–48, 47*t*
Benue River and Basin
   hydrography, 17–18
   hydrology, 44–45
   map, 76*m*
   TSS, 52, 53*f*
Burkina Faso, 2–3*t*, 4, 46–48, 47*t*

Cameroon, 2–3*t*, 4–5, 46–48, 47*t*
Chad, 2–3*t*, 5, 46–48, 47*t*
civil war and conflict, effects of, 66
climate, 25–29, 25*t*, 26*f*, 28*f*, 29*f*, 86*t*, 90*f*
colonial era, Niger River basin during, 8
cooperative development of Niger River basin, xiii–xiv, 58–69
   data for decision-making, 65, 84*m*, 127–128
   dynamic approach, need for, 60
   environmental factors, 66–67
   factors involved in, 63–67
   goals of, 58–59
   historical background, 7–10
   institutional and legal foundations, 60–62
   international partners in, 69
   key ingredients for success of, 68–69
   moving from unilateral development to, 59–60, 69
   SDAP (Sustainable Development Action Program), xiv, 10, 61, 63–64, 69
   Shared Vision process, xiv, 10, 61, 63–64, 69
   sociological factors, 65–67
   types of cooperation, 64, 65*f*
Côte d'Ivoire, 2–3*t*, 5, 46–48, 47*t*

data management, 65, 84*m*, 127–128
debt sustainability, 65
definitions, 129–130
degradation, environmental
   reasons for, 66–67
   water quality, 56–57
Delta. *See* Inland Delta; Lower Niger and Delta
development aims. *See* cooperative development of Niger River basin
dissolved solids, total (TDS), 53–56, 55*f*, 56*t*, 124*f*, 126*f*, 130

environment and ecology, 24, 66–67. *See also* physical characteristics of Niger River basin

environmental degradation
    reasons for, 66–67
    water quality, 56–57
evaporation rates, 37–41, 38f, 39f,
    39t, 40f, 83m, 87t.89f

flooding
    in Inland Delta and Lakes
        District, 41–42f
    in Middle Niger, 43f
    in Upper Basin, 35–36, 35t
flow rates, 33t, 37t, 40f, 47t, 49t,
    51t, 94–123t
FONDAS (Niger River Basin
    Development Fund), 9

geography. *See* physical
    characteristics of Niger
    River basis
geology and hydrogeology, 20–22
glossary, 129–130
Guinea, 2–3t, 5–6, 33t, 46–48, 47t

Heads of State Summit, xiv, 10, 61,
    63, 68
humidity, 89f, 90f
hydrogeology, 20–22
hydrography, 12–19
hydrology, 30–48
    Bani Watershed, 34–36
    Benue River and Basin, 44–45
    country-specific features,
        46–48, 47t
    evaporation rates, 37–41, 38f,
        39f, 39t, 40f, 83m, 87t.89f
    flow rates, 33t, 37t, 40f, 47t, 49t,
        51t, 94–123t
    humidity, 89f, 90f
    Inland Delta and Lakes District,
        36–42, 37t, 38f, 39f, 39t,
        40–42f
    Lower Niger and Delta, 45–46
    maps, 71–85m
    Middle Niger, 42–44, 43f

monitoring system, 65, 84m,
    127–128
parameters, 35t, 91t, 93t
rainfall, 27–29, 28f, 29f, 81–83m,
    88f, 92t
runoff, 28, 29f
Upper Basin, 32–36, 33t, 35t
HYDRONIGER monitoring
    system, 65, 84m, 127–128

infrastructure sharing, 64–65
Inland Delta and Lakes District
    evaporation in, 37–41, 38f, 39f,
        39t, 40f
    flooding in, 41–42f
    hydrogeology, 21
    hydrography, 14–15
    hydrology, 36–42, 37t, 38f, 39f,
        39t, 40–42f
    map, 73m, 85m
    TDS, 54, 55f
    TSS, 50–52, 51t
international partners in
    cooperative development, 69
ionic composition, 54, 56t, 126f
isohyets, 27, 28f, 81–82m, 129

Lakes District. *See* Inland Delta
    and Lakes District
legal framework for cooperative
    development, 62
Lower Niger and Delta
    hydrogeology, 21–22
    hydrography, 18–19, 31f
    hydrology, 45–46
    map, 77m
    TDS, 54–56, 56t
    TSS, 52, 53f

Mali, 2–3t, 6, 46–48, 47t
maps, 71–85m
master plans, limitations of, 60
Middle Niger
    climate, 86t

INDEX 143

hydrography, 15–17
hydrology, 42–44, 43f
maps, 74–75m
TDS, 54–56, 56t
TSS, 52, 53f, 54t
migration flows, effect of, 66
monitoring system, 65, 84m, 127–128

natural environment, 24, 66–67. *See also* physical characteristics of Niger River basin
navigation
  map of navigable segments, 80m
  physical characteristics of navigable segments, 19–20
  TSS affecting, 52–53
NBA. *See* Niger River Basin Authority
Niger, 2–3t, 6–7, 46–48, 47t
Niger Basin Convention of 1980, 9, 60
Niger Delta. *See* Inland Delta; Lower Niger and Delta
Niger River basin, vii–viii, x–xi, 1
  cooperative development (*See* cooperative development of Niger River basin)
  countries of, 1–7, 2–3t
  history of, 7–10
  maps, 71–85m
  origins of name, 1
  physical characteristics (*See* physical characteristics of Niger River basin)
Niger River Basin Authority (NBA), xiii–xiv
  audit of, 63
  historical background and establishment of, 9–10
  institutional foundation for cooperative development and, 60–61
  potential roles of, 67
  renewal of, 69
Niger River Basin Development Fund (FONDAS), 9
Nigeria, 2–3t, 7, 46–48, 47t

OMVS (Senegal River Basin Organization), 64

physical characteristics of Niger River basin, xi–xii, 1, 11–29
  climate, 25–29, 25t, 26f, 28f, 29f, 86t, 90f
  flow rates, 33t, 37t, 40f, 47t, 49t, 51t, 94–123t
  general physical environment, 1, 11–12
  geology and hydrogeology, 20–22
  hydrography, 12–19
  hydrology (*See* hydrology)
  maps, 71–85m
  natural environment and ecosystem, 24
  navigable segments, 19–20
  rainfall, 27–29, 28f, 29f, 81–83m, 88f, 92t
  runoff, 28, 29f
  soil types, 22–23, 23f, 129–130
poverty reduction as aim of cooperative development, 58–59

quality of water, 56–57

rainfall, 27–29, 28f, 29f, 81–83m, 88f, 92t. *See also* hydrology
runoff, 28, 29f

SDAP (Sustainable Development Action Program), xiv, 61, 63–64, 69
Senegal River Basin Organization (OMVS), 64

Shared Vision process, xiv, 10, 61, 63–64, 69
socioeconomic characteristics of Niger River basin countries, 23–24*t*
sociological factors in cooperative development, 65–67
soil types, 22–23, 23*f*, 129–130
solids
  total dissolved (TDS), 53–56, 55*f*, 56*t*, 124*f*, 126*f*, 130
  total suspended (TSS), 48–53, 49*f*, 50–51*t*, 53*f*, 54*t*, 125*f*, 130
subsidiarity, xiv, 67, 131n10
subsidiary agreements, need for, 62
suspended solids, total (TSS), 48–53, 49*f*, 50–51*t*, 53*f*, 54*t*, 125*f*, 130
Sustainable Development Action Program (SDAP), xiv, 61, 63–64, 69

temperature, seasonal variations in, 90*f*. *See also* climate
total dissolved solids (TDS), 53–56, 55*f*, 56*t*, 124*f*, 126*f*, 130
total suspended solids (TSS), 48–53, 49*f*, 50–51*t*, 53*f*, 54*t*, 125*f*, 130

Upper Basin. *See also* Bani Watershed
  climate, 86*t*
  flooding in, 35–36, 35*t*
  hydrogeology, 20–21
  hydrography, 12–14
  hydrology, 32–36, 33*t*, 35*t*
  map, 72*m*
  seasonal variability in, 34–35
  TDS, 53–54, 55*f*
  TSS, 48–50, 49*f*, 50*t*, 55*f*

war and conflict, effects of, 66
water resources, xii–xiii, 30–57
  hydrogeology, 20–22
  hydrography, 12–19
  hydrology (*See* hydrology)
  quality of water, 56–57
  rainfall, 27–29, 28*f*, 29*f*, 81–83*m*, 88*f*, 92*t*
  runoff, 28, 29*f*
  TDS, 53–56, 55*f*, 56*t*, 124*f*, 126*f*, 130
  TSS, 48–53, 49*f*, 50–51*t*, 53*f*, 54*t*, 125*f*, 130
World Bank as international partner in cooperative development, 69

Eco-Audit
Environmental Benefits Statement

The World Bank is committed to preserving endangered forests and natural resources. We have chosen to print The Niger River Basin: A Vision for Sustainable Management on 30% post-consumer recycled fiber paper, processed chlorine free. The World Bank has formally agreed to follow the recommended standards for paper usage set by the Green Press Initiative—a nonprofit program supporting publishers in using fiber that is not sourced from endangered forests. For more information, visit www.greenpressinitiative.org.

In 2005, the printing of these books on recycled paper saved the following:

Trees*	Solid Waste	Water	Net Greenhouse Gases	Electricity
20	330	2,991	648	1,202
*40' in height and 6-8" in diameter	Pounds	Gallons	Pounds	KWH